Tobias Meier

High-Power CW Green Lasers for Optical Metrology

Tobias Meier

High-Power CW Green Lasers for Optical Metrology

and Their Joint Benefit in Particle Physics Experiments

Südwestdeutscher Verlag für Hochschulschriften

Impressum/Imprint (nur für Deutschland/only for Germany)
Bibliografische Information der Deutschen Nationalbibliothek: Die Deutsche Nationalbibliothek verzeichnet diese Publikation in der Deutschen Nationalbibliografie; detaillierte bibliografische Daten sind im Internet über http://dnb.d-nb.de abrufbar.
Alle in diesem Buch genannten Marken und Produktnamen unterliegen warenzeichen-, marken- oder patentrechtlichem Schutz bzw. sind Warenzeichen oder eingetragene Warenzeichen der jeweiligen Inhaber. Die Wiedergabe von Marken, Produktnamen, Gebrauchsnamen, Handelsnamen, Warenbezeichnungen u.s.w. in diesem Werk berechtigt auch ohne besondere Kennzeichnung nicht zu der Annahme, dass solche Namen im Sinne der Warenzeichen- und Markenschutzgesetzgebung als frei zu betrachten wären und daher von jedermann benutzt werden dürften.

Coverbild: www.ingimage.com

Verlag: Südwestdeutscher Verlag für Hochschulschriften GmbH & Co. KG
Heinrich-Böcking-Str. 6-8, 66121 Saarbrücken, Deutschland
Telefon +49 681 37 20 271-1, Telefax +49 681 37 20 271-0
Email: info@svh-verlag.de

Approved by: Hannover, Leibniz Universität, Dissertation, 2011

Herstellung in Deutschland:
Schaltungsdienst Lange o.H.G., Berlin
Books on Demand GmbH, Norderstedt
Reha GmbH, Saarbrücken
Amazon Distribution GmbH, Leipzig
ISBN: 978-3-8381-1434-7

Imprint (only for USA, GB)
Bibliographic information published by the Deutsche Nationalbibliothek: The Deutsche Nationalbibliothek lists this publication in the Deutsche Nationalbibliografie; detailed bibliographic data are available in the Internet at http://dnb.d-nb.de.
Any brand names and product names mentioned in this book are subject to trademark, brand or patent protection and are trademarks or registered trademarks of their respective holders. The use of brand names, product names, common names, trade names, product descriptions etc. even without a particular marking in this works is in no way to be construed to mean that such names may be regarded as unrestricted in respect of trademark and brand protection legislation and could thus be used by anyone.

Cover image: www.ingimage.com

Publisher: Südwestdeutscher Verlag für Hochschulschriften GmbH & Co. KG
Heinrich-Böcking-Str. 6-8, 66121 Saarbrücken, Germany
Phone +49 681 37 20 271-1, Fax +49 681 37 20 271-0
Email: info@svh-verlag.de

Printed in the U.S.A.
Printed in the U.K. by (see last page)
ISBN: 978-3-8381-1434-7

Copyright © 2011 by the author and Südwestdeutscher Verlag für Hochschulschriften GmbH & Co. KG and licensors
All rights reserved. Saarbrücken 2011

Contents

1. Introduction   1

2. High-power 532 nm single-frequency TEM$_{00}$ laser sources   7
    - 2.1. Light propagation through media 9
        - 2.1.1. Propagation in a linear medium 10
        - 2.1.2. Gaussian beams and plane waves 11
        - 2.1.3. Optical resonators 14
        - 2.1.4. Propagation in a nonlinear medium 17
        - 2.1.5. Nonlinear susceptibility 20
        - 2.1.6. (Quasi-)Phase matching 21
        - 2.1.7. The coupled equations of motion for SHG 23
    - 2.2. Modelling high-power second harmonic generation 25
        - 2.2.1. The Boyd-Kleinman integral 25
        - 2.2.2. Approximate analytic solutions 27
        - 2.2.3. Doubling schemes and suitable crystals 28
        - 2.2.4. Processes that lead to thermal dephasing 29
        - 2.2.5. Models with and without thermal dephasing 32
    - 2.3. Converting an intermediate power metrology laser 39
        - 2.3.1. Reported green light power levels from PPKTP 40

		2.3.2. General external-cavity design considerations 40
		2.3.3. Design of the resonant PPKTP SHG stage 42
		2.3.4. Experimental setup 46
		2.3.5. Results and discussion 50
	2.4.	A 130 W CW single-frequency TEM$_{00}$ green laser source 66
		2.4.1. Highest reported harmonic power levels for green light 67
		2.4.2. Choice of design and crystal 69
		2.4.3. Design of the SHG experiment 69
		2.4.4. Experimental setup 74
		2.4.5. Conventional analysis of transverse mode structure 78
		2.4.6. Reduction of transverse mode analysis uncertainty 81
		2.4.7. Results and discussion of the SHG experiment 83
	2.5.	Summary and outlook 100

3. ALPS I project - Particle physics with high-power green light 103

	3.1.	Particle physics with laser light 105
		3.1.1. The standard model of particle physics 105
		3.1.2. Virtual particles and Feynman diagrams 108
		3.1.3. One-loop process magnetic vacuum birefringence 109
		3.1.4. Particle accelerators and LSW experiments 110
		3.1.5. Problems of the Standard Model 112
	3.2.	Hypothetical hidden sector particles coupling to photons 116
		3.2.1. Three kinds of hypothetical particles 116
		3.2.2. WISPs as answers to puzzling astrophysical observations . . . 120
	3.3.	State of the art of LSW experiments in the world 121
	3.4.	The first LSW experiment with production resonator 123
		3.4.1. Design and experimental setup of ALPS I (phase 1) 125
		3.4.2. Results of ALPS I (phase 1) and discussion 140
	3.5.	ALPS I (phase 2) - the world's most sensitive WISP detector 151
		3.5.1. Experimental setup of ALPS I (phase 2) experiment 152
		3.5.2. Results of ALPS I (phase 2) and discussion 157

   3.6. Summary and outlook 170

4. ALPS II - High-precision optical metrology boosts sensitivity   **173**
   4.1. Signal enhancement on the regeneration side 174
        4.1.1. Local oscillator 174
        4.1.2. Regeneration cavity 176
        4.1.3. Possible experimental realizations 177
   4.2. Basic design study for the ALPS II experiment 178
        4.2.1. Cavity design 178
        4.2.2. Power buildup factors 179
        4.2.3. Cavity length and free apertures 180
        4.2.4. Highest intensity 181
        4.2.5. Spatial overlap of cavity modes 182
        4.2.6. Mirror lifetime 185
   4.3. A proposed experimental setup 187
        4.3.1. Proposed experimental setup for ALPS II 187
        4.3.2. Technical challenges 191
   4.4. Projected sensitivity 192
   4.5. Summary and outlook 194

5. Conclusion   **197**

A. Basic optics   **203**
   A.1. Maxwell's equations 203
   A.2. General solution describing light propagation 204
   A.3. Intensity and power 205

Bibliography   **207**

Acknowledgements   **225**

List of abbreviations

ALPS	Any Light Particle Search
AM	amplitude modulation
AOM	acousto-optical modulator
AP	axion-like particle
BBO	beta barium borate (β-BaB_2O_4)
BiBO	bismuth triborate (BiB_3O_6)
CCD	charge-coupled device
CMB	cosmic microwave background
CW	continuous-wave
DESY	Deutsches Elektronen Synchrotron Hamburg, Germany
DFG	difference frequency generation
EOM	electro-optical modulator
GRIIRA	green-induced infrared absorption
HEPA	high efficiency particulate air
HERA	Hadron-Electron Ring Accelerator
IBS	ion-beam sputtering
KTP	potassium titanyl phosphate ($KTiOPO_4$)
LBO	lithium triborate (LiB_3O_5)
LG	transverse electromagnetic Laguerre Gauss
LHC	Large Hadron Collider

LIGO	Laser Interferometer Gravitational-Wave Observatory
LN	lithium niobate (LiNbO$_3$)
LO	local oscillator
LSD	linear spectral density
LSW	light shining through a wall
LZH	Laser Zentrum Hannover
MCP	mini-charged particle
MHP	massive hidden sector photon
MOPA	master-oscillator power amplifier system
NIR	near-infrared
NPRO	non-planar ring oscillator
PDH	Pound-Drever-Hall
PID	proportional-integral-derivative
PM	phase modulation
PPKTP	periodically-poled KTP
PPLN	periodically poled LN
PPSLT	periodically poled stoichiometric lithium tantalate (LiTaO$_3$)
PSL	Pre-Stabilised Laser
QCD	quantum chromodynamics
RMS	root mean square
ROC	radius of curvature
SFG	sum frequency generation
SHG	second harmonic generation
SLT	stoichiometric lithium tantalate (LiTaO$_3$)
SM	Standard Model of particle physics
SNR	signal-to-noise ratio
TEM	transverse electromagnetic Hermite Gauss
UGF	unity gain frequency

WISP weakly interacting slim (or sub-eV) particle
WMAP Wilkinson Microwave Anisotropy Probe

List of symbols

α	combined power absorption coefficient; ⋆
α_l	linear power absorption coefficient; ⋆
δT_bw	temperature acceptance bandwidth of nonlinear crystal
Δk	wave vector difference due to phase mismatch
ϵ_0	vacuum permittivity (a.k.a. electric constant)
ϵ_r	relative permittivity (a.k.a. dielectric constant)
η_{00}	power fraction contained in TEM_{00} mode
η_SHG	external SHG conversion efficiency
κ_th	thermal conductivity
λ_0	laser wavelength in vacuum; ⋆
μ_0	vacuum permeability (a.k.a. magnetic constant)
μ_r	relative permeability
ξ	Boyd-Kleinman focussing parameter
σ_opt	optimized dephasing parameter of Boyd-Kleinman integral
χ	coupling constant of photon field to field of massive hidden sector photon
$\chi_\text{e}^{(1)}$	linear electric susceptibility
$\chi_\text{e}^{(2)}$	nonlinear electric susceptibility of second order
$\chi_\text{m}^{(1)}$	linear magnetic susceptibility

A	summed fractional round trip losses of cavity
A_m	fractional round trip losses of cavity due to mirrors
A_p	summed passive fractional round trip losses of cavity
$A_\mathrm{p,a}$	fractional round trip losses due to absorption
$A_\mathrm{p,s}$	fractional round trip losses due to scattering
A_SHG	fractional round trip losses due to SHG
c_0	speed of light in vacuum
d	geometric distance
d_eff	effective nonlinearity of crystal
E	rapidly varying electric field
$\mathcal{F}$	cavity finesse
FSR	free spectral range of cavity
$FWHM$	resonator FWHM linewidth
g_-	coupling constant of photon field to field of pseudoscalar axion-like particles
g_+	coupling constant of photon field to field of scalar axion-like particles
$\hat{h}$	Boyd-Kleinman integral
I	optical intensity; $\star$
I_w	peak intensity circulating inside SHG resonator at waist inside nonlinear crystal; $\star$
$\tilde{J}$	Boyd-Kleinman integrand
k	wave vector of laser light; $\star$
L_k	geometric crystal length
L_m	geometric length of magnetic field in production and regeneration region
L_rt	optical round trip length
n	real refractive index; $\star$

N_p	number of photons in production region
N_r	number of photons in regeneration region
$\mathcal{P}$	general probability
P	optical power; $\star$
$\tilde{P}$	rapidly varying atomic polarization of medium
P_{circ}	power circulating inside cavity (single direction)
P_{inc}	incident power, often on cavity input coupler; $\star$
P_{loss}	power lost inside cavity
P_{refl}	power reflected from cavity input coupler
P_{trans}	power transmitted through cavity output coupler
PB	power buildup factor of cavity
PB_{max}	resonant power buildup factor of cavity
PB_p	power buildup factor of production cavity
PB_r	power buildup factor of regeneration cavity
T_{dc}	harmonic power transmission of dichroic mirror
T_{in}	power transmission of cavity input coupler
T_{out}	power transmission of cavity output coupler
U	general slowly varying complex field amplitude
$\check{U}$	slowly varying complex amplitude of an arbitrary Gaussian beam
U_0	slowly varying complex amplitude of a TEM_{00} beam; $\star$
V_{nm}	complex amplitude of a transverse electromagnetic mode
w_0	laser beam radius at focus (a.k.a. waist); $\star$
z_r	Rayleigh range
$\star$	subscripts 'f' for fundamental light and 'h' for harmonic light

Introduction

The realization of the first laser by Maiman in the year 1960 started a new era in physics [91]. One example is the first observation of the effect of second harmonic generation (SHG) by Franken et al. only one year later [47].

The collimated beam, which is emitted from a laser, enables the achievement of high optical intensities. Furthermore, the good spatial and temporal coherence properties of this beam allows the use of its wavelength as reference for distance measurements. This increased the precision of such measurements by orders of magnitude. The effect of SHG in turn helped to realize laser sources for a large number of different wavelengths.

Today, interferometric gravitational wave detectors measure the phase difference between two beams to detect distance changes between two mirrors [150]. If a gravitational wave travels through the detector it will change this difference. These waves represent disturbances of spacetime and were predicted by Albert Einstein in 1916 in the context of his theory of general relativity [126].

The hunt for the first direct detection of these waves is still ongoing, as the relative length changes caused by gravitational waves on Earth are tiny. Even for the most rapidly accelerated and most massive objects in the universe they are expected to be on the order of only 10^{-22} or less [127]. Therefore enormous efforts were made and

1. Introduction

are still underway to allow for a direct detection in the near future. These efforts emerged into the realization of several large-scale gravitational wave detectors (e.g. GEO 600 [150]) and into future missions being under construction (e.g. Advanced LIGO [50]) and in planning phase (e.g. ET [116] or DECIGO [52]). For all of them continuous-wave (CW) sources of laser light with highest stability and beam quality are required. Especially, these lasers have to emit a high power at a single frequency [149].

Once the sensitivity of those future detectors is good enough to detect gravitational waves on a regular basis, the scientific gain will also be enormous. Gravitational wave observatories will open up a completely new perspective to observe astrophysical processes, fundamentally different from everything, which is accessible to date. From this new perspective scientists will learn a lot about our universe from its current state back to its very beginning [127].

Another way to learn a lot about our universe is the analysis of the constituents of matter and the fundamental forces. Apart from gravity, these aspects are described accurately today by the Standard Model of particle physics (SM) [113]. However, the SM cannot be a complete description of our universe. This is obvious as it excludes gravity. Additionally, it misses explanations for the occurrence of dark matter [160] and dark energy [109], whose existence was approved by the seven-year results of the Wilkinson Microwave Anisotropy Probe (WMAP) satellite mission [70].

In contrast, string theory and other extensions of the SM are able to include those effects [113]. Usually these candidates for extension or substitution of the SM predict a large set of light particles, which interact only very feebly with ordinary matter [68]. Such a particle is generally denoted as weakly interacting slim (or sub-eV) particle (WISP). Additionally, several puzzling astrophysical observations can be understood much better, if WISPs exist [87]. The same is true for the non-observation of CP symmetry breaking in case of the strong force [42]. However, up to now such WISPs could not be observed by any laboratory or collider experiment. In fact, this might be very well caused by the simple fact that they are weakly-interacting and thus evade detection by collider experiments due to a too low primary particle flux.

Interestingly, several WISP species are expected to couple also to photons as their mass is low enough to allow production from photon-photon interactions [68]. This concept is utilized in light shining through a wall (LSW) experiments, which search for the oscillation of photons from a light beam into WISPs and vice versa [4]. In these experiments the primary particle flux can be made extraordinary large as, for instance, a beam of green light with an optical power of 100 W corresponds to a huge flux of photons of $3 \times 10^{20} \, \frac{1}{s}$.

In addition to high-power light sources, LSW experiments require extremely low-noise detectors, and large-scale vacuum systems and strong magnetic fields with lengths of ten meters and more are necessary. To exploit the full length, the light has to be coherent. Additionally, the usually small apertures of the magnets require a confined beam. Thus the application of lasers seems inevitable. Furthermore, in this thesis optical resonators were used to further increase the primary particle flux. The application of this optical metrology technique even requires a CW single-frequency laser source with good beam quality and stability.

If WISPs are found they will be a direct proof for the existence of physics beyond the Standard Model. The knowledge about their precise mass and coupling to photons would severely constrain Standard Model extensions or even falsify some of them. This also would fundamentally change our understanding of the universe from star evolution to the Big Bang, as completely new constituents and interactions would have to be taken into account [118].

Both examples given above require high-power CW single-frequency metrology laser sources with good beam quality. Such sources have already been constructed in the infrared spectral region [151]. As in both cases the performance of the experiment is at least theoretically improved with decreasing wavelength of the applied laser, it is desirable to have such a laser also available in the green visible spectral region.

The first part of this thesis deals with the construction of such a high-power CW single-frequency laser source with a wavelength of 532 nm. The second part deals with the fundamental optimization of LSW experiments, conducted by the author as member of the Any Light Particle Search (ALPS) collaboration. The optimization

1. Introduction

was done by the application of optical resonators and of a laser source with the previously mentioned properties. Implementation of this concept into the ALPS experiment was the crucial step to achieve large sensitivity enhancements.

Outline of this Thesis

The structure of this thesis is described in the following.

Chapter 2 deals with the construction of high-power single-frequency laser sources for the green visible spectral region via the effect of SHG. Starting from the theoretical basics, a model for SHG in an external cavity is developed. Then two experiments with essentially different nonlinear crystals are described. In both cases optimum single-pass conversion was intentionally avoided as suggested by the modelling process, and both cases achieved leading results. The result was a laser source, which met all basic requirements imposed on a laser source for a gravitational wave detector, and which emitted an unprecedented power of 134 W at 532 nm! A new measurement technique was developed to reduce the uncertainty of conventional spatial beam quality measurements.

Chapter 3 presents the ALPS I experiment. This is the first LSW experiment world-wide, which utilizes a CW laser in combination with an optical resonator comprising the WISP production region. After explaining the basics of LSW experiments and related particle physics, the WISP types of interest are introduced and motivations are given to search for them. Then the two phases of the ALPS I experiment are described. The first phase demonstrated the compatibility of the long optical resonator with an existing large-scale LSW experiment. In the second phase the experiment was upgraded and subsequently achieved the highest sensitivity world-wide.

In Chapter 4, design considerations for a future ALPS II experiment are described, which is currently in its planning and preparation phase. This experiment will show an impressive enhancement in sensitivity compared to its precursor, as it will incorporate a so-called regeneration cavity. First, this and other key techniques for the sensitivity enhancement are explained. Then, basic design considerations for an unprecedented specific realistic implementation are presented, and its theoretically achievable sensitivity is estimated. The latter will be orders of magnitude higher than that of ALPS I!

Finally, in Chapter 5, the most important conclusions of this thesis together with their impact and implications are summarized.

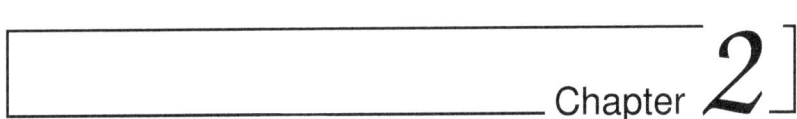

High-power 532 nm single-frequency TEM$_{00}$ laser sources

High-power CW green light is important in many applications of science and engineering. Examples are pumping of titanium-sapphire lasers or dye lasers [17], the latter enabling the construction of laser guide star adaptive optics systems, which become increasingly important for astronomy [117]. By the process of SHG the generation of CW deep-UV radiation from green light is possible, which is utilized in fiber Bragg grating production or semiconductor mask inspection [123].

Another application of green laser radiation can be found in the field of laser material processing, because CW-(micro-)welding in the green spectral region is advantageous to the usual application of infrared lasers when materials are processed, that have significantly higher reflectivities for infrared light like copper or gold [65]. A low amount of power in higher-order transversal modes increases micro-welding precision.

Future gravitational-wave detectors like the Einstein Telescope (ET) might benefit from a single-mode high-power laser source at 532 nm [52]. Such a detector main-

2. High-power 532 nm single-frequency TEM$_{00}$ laser sources

tains the same signal-to-quantum-noise ratio if both the optical wavelength and the circulating optical power are halved [23]. However, thermal effects would be greatly reduced. In any case the signal strength grows linearly with the inverse wavelength. Plans for future space-borne missions like DECIGO or BBO already incorporate this idea [52]. As coatings and optics have to be tested for quality and reliability at the wavelengths applied in these missions, a high-power CW laser source at 532 nm might become a valuable tool.

Finally, so-called light shining through a wall (LSW) experiments in particle physics need powerful lasers as particle sources. A wavelength of 532 nm would in this case have two advantages, namely the possibility for longer conversion regions and the availability of more sensitive detectors. Chapters 3 and 4 of this thesis will deal with such experiments in more detail.

Compared to argon ion gas laser systems, neodymium doped solid-state lasers emitting at 1064 nm are much more efficient. Thus it is a promising technique to generate high-power laser light at 532 nm via the process of SHG from such an infrared laser.

A very large amount of scientific publications deal with SHG of light with a wavelength of 532 nm, some of which could demonstrate extraordinary high conversion efficiencies of up to 92 % at low power levels on the order of 100 mW [141, 142]. Such efficiencies are indeed remarkable as one has to keep in mind that in many other fields of research such higher-order nonlinear effects are completely neglected in everyday perturbation analysis.

Considerably less publications deal with the generation of harmonic power levels in the multi-watt or even 100 W regime. At such power levels the single-pass conversion efficiencies of the nonlinear crystals are often strongly degraded by thermal dephasing processes originating from optical absorption. Some contributing effects to the overall absorption coefficients are intensity dependent.

In the scope of this thesis a simple model was developed, which showed that the scheme of external-cavity doubling is beneficial for high-power SHG. It allows to strongly reduce the intensities of the fundamental and harmonic beam and correspondingly the intensity dependent absorption effects without sacrificing the external conversion efficiency. Based on this idea a resonant high-power SHG stage with a high nonlinearity crystal was set up, which achieved the highest long-term stable single-frequency harmonic power reported so far and known to the author for this material and wavelength. Its performance was in good agreement with the developed model but was still clearly limited by thermal dephasing processes.

In a next step a very high power metrology infrared laser provided the fundamental beam. It emitted a power of 150 W. As thermal dephasing could not be avoided in the previous setup the strategy was changed. In contrast to the previous experiment, here a *low* nonlinearity crystal with proven resilience against high optical powers was utilized. In this experiment an extraordinary high single-frequency harmonic power of 134 W could be achieved in good agreement with the model mentioned above! This is, to the best of the author's knowledge, the highest scientifically published single-frequency power so far for a 532 nm wavelength. Furthermore, a technique was developed, which allowed to reduce the uncertainty in the result for the fundamental transversal mode content of the harmonic power.

In Sec. 2.1 of this chapter the basics of light propagation and second harmonic generation are explained. These are used in Sec. 2.2 to develop a simple model for high-power SHG devices, which incorporates thermal dephasing processes to some extend. Sec. 2.3 reports on the conversion experiment with the high-nonlinearity crystal and its comparison with the model. Finally, Sec. 2.4 describes the improved technique to determine the fundamental transversal mode content, and it presents the extraordinary high harmonic power SHG experiment with the low-nonlinearity crystal.

Some results on an essentially identical experiment like the one described in Section 2.3 have already been published in [97] and the main results of Section 2.4 can be found in the literature as well [98].

2.1. Light propagation through media

In this subsection the basic models and effects of light propagation, optical resonators and second harmonic generation are reviewed in the extend necessary for

2. High-power 532 nm single-frequency TEM$_{00}$ laser sources

the contents of this thesis.

2.1.1. Propagation in a linear medium

Light propagation is governed by Maxwell's equations. From them a simplified one-dimensional differential equation can be derived, which describes the propagation of light in a rather general medium or vacuum (see A.1 in appendix) [14, 105, 40]

$$\nabla^2 E(z,t) - \frac{1+\chi_m^{(1)}}{c_0^2}\frac{\partial^2}{\partial t^2}E(z,t) = \frac{1+\chi_m^{(1)}}{\epsilon_0 c_0^2}\frac{\partial^2}{\partial t^2}\tilde{P}(z,t) \quad . \tag{2.1}$$

Here E and $\tilde{P}$ are the rapidly varying electric field and atomic polarization field, $\chi_m^{(1)}$ is the magnetic susceptibility, c_0 is the speed of light in vacuum and ϵ_0 the electric constant. All parameters are in SI units.

The right hand side of Eq. (2.1) represents a source term for the propagating electric field. If the wave propagates in vacuum this term would be assumed to be zero (because their is nothing to polarize) indicating that the complex amplitude of the wave is not changed during propagation. But obviously this term is nonzero if the light propagates inside a medium. Thus one would expect some change of the amplitude and/or phase of the electric field. This is indeed the case and can be seen if one inserts the explicit expression for $\tilde{P}$ in the case of propagation inside an ideal *linear* medium [39, 105]

$$\tilde{P}(t) = \epsilon_0 \chi_e^{(1)}(\nu) E(t) \quad , \tag{2.2}$$

$$\Rightarrow \nabla^2 E(\nu,t) - \frac{(1+\chi_e^{(1)}(\nu))(1+\chi_m^{(1)})}{c_0^2}\frac{\partial^2}{\partial t^2}E(\nu,t) = 0 \quad ,$$

where the relative permittivity $\epsilon_r(\nu) = 1 + \chi_e^{(1)}(\nu)$ and relative permeability $\mu_r = 1 + \chi_m^{(1)}$ are complex parameters that depend on the optical frequency and in birefringent crystals also on the orientation of the optical polarization and propagation direction relative to the crystal's optical axes. Their real parts have the ability to change the phase of the wave and their imaginary parts change the wave's amplitude. They are related to the real quantities refractive index $\mathbf{n}(\nu)$ and extinction coefficient $\kappa(\nu)$ by $\epsilon_r \mu_r = (\mathbf{n} + i\kappa)^2$.

2.1. Light propagation through media

After this reformulation the source term in the above equation has vanished showing that within a linear medium no other effects are to be expected.

2.1.2. Gaussian beams and plane waves

The most general solution of Eq. (2.2), which is able to describe electro-magnetic waves, is given by

$$E(z,t) = \frac{1}{2} U(x,y,z,t)\, e^{i(kz - 2\pi\nu t)} + (cc) \quad . \tag{2.3}$$

is an unnecessarily general model for a beam of laser light. To find a suitable analytic expression for a realistic laser beam one claims the complex amplitude U in Eq. (2.3) to be less general. The corresponding solutions are called Gaussian beams. Their slowly varying complex amplitude has the general form [129]

$$\check{U}(x,y,z,t) = U_0(z,t) \sum_{m,n} C_{nm}(z,t)\, V_{nm}(x,y,z) \quad , \tag{2.4}$$

$$V_{nm}(x,y,z) = \frac{1}{\sqrt{2^{n+m-1}\, n!\, m!\, \pi}\, w(z)} H_n\left[\frac{\sqrt{2}}{w(z)} x\right] H_m\left[\frac{\sqrt{2}}{w(z)} y\right]$$
$$\times \exp\left[-\frac{r^2}{w(z)^2}\right] \exp\left\{i\left[\frac{k}{2R(z)} r^2 - (n+m+1)\psi(z)\right]\right\} \quad ,$$

with $H_n[x]$ denoting a Hermite polynomial of order n, the radial distance $r = \sqrt{x^2 + y^2}$ and the beam radius $w(z)$, phase front radius of curvature $R(z)$ and Gouy phase $\psi(z)$ given by

$$w(z) = w_0 \sqrt{1 + \left(\frac{z}{z_r}\right)^2} \quad , \tag{2.5}$$

$$R(z) = z\left[1 + \left(\frac{z_r}{z}\right)^2\right] \quad ,$$

$$\psi(z) = \arctan\left(\frac{z}{z_r}\right) \quad ,$$

and w_0 denotes the beam's nonzero radius at its focus (i.e. its narrowest point, also called beam waist). While the sign of the phase in Eq. (2.3) was chosen by

2. High-power 532 nm single-frequency TEM$_{00}$ laser sources

convention it is important to choose the signs of the phase contributions in Eq. (2.4) in the correct relative relation to this convention.

Obviously the radius of a Gaussian beam necessarily grows with increasing distance from its focus and the divergence angle depends on the wavelength and waist. This effect is called diffraction. A typical distance is the Rayleigh range, which is defined as

$$z_r = \frac{\pi n w_0^2}{\lambda_0} \quad .$$

It denotes the distance from the waist, at which the beam radius has increased by a factor $\sqrt{2}$. Hence the intensity of a Gaussian beam also depends on the distance from its waist, which will become important for the intensity dependent nonlinear effects discussed below.

The transversally varying part of a Gaussian beam (i.e. V$_{nm}$) is often called transverse electromagnetic Hermite Gauss (TEM) mode. These TEM modes are normalized such that the surface integral over the cross-sectional plane of their squared modulus gives unity

$$\int_0^\infty r \int_0^{2\pi} |V_{nm}|^2 \, d\phi \, dr = 1 \quad .$$

The set of TEM$_{nm}$ modes made up of all non-negative integers n and m form an orthonormal basis. This means on the one hand that the projection of any mode onto any other results in an orthonormality relation

$$\delta_{nj} \delta_{mk} = \int_{-\infty}^{\infty} \int_{-\infty}^{\infty} V_{nm}(x,y,z) \, [V_{jk}(x,y,z)]^\dagger \, dx \, dy \quad . \tag{2.6}$$

On the other hand an arbitrary laser beam with a given propagation direction can always be expressed as a superposition of TEM modes. Such a decomposition is performed by calculating the overlap integral of the reference transverse mode V$_{nm}$ and the beam to be decomposed $\check{U}_d(x,y,z,t)$ [75]

$$U_0 C_{nm} = \int_{-\infty}^{\infty} \int_{-\infty}^{\infty} \check{U}_d(x,y,z,t) \, [V_{nm}(x,y,z)]^\dagger \, dx \, dy \quad . \tag{2.7}$$

2.1. Light propagation through media

The transverse mode coefficients C_{nm} then give the power fraction $|C_{nm}|^2$ and the electric field fraction $|C_{nm}|$ of $\check{U}_d$ contained within V_{nm}. The phase shift of the corresponding transversal mode component of $\check{U}_d$ relative to V_{nm} is given by $\arg(C_{nm})$. Energy conservation then results in a normalization condition with the beam's power P_d (for the general definition of intensity and optical power in this thesis see A.3 in the appendix)

$$|U_0|^2 \sum_{m,n} |C_{nm}(z,t)|^2 = \int_0^\infty r \int_0^{2\pi} |\check{U}_d|^2 \, d\phi \, dr = \frac{2}{n\epsilon_0 c_0} P_d \quad .$$

The radius and divergence angle of a Gaussian beam grows with a decreasing relative fraction of its power contained inside its fundamental mode TEM_{00} [129].

Generally speaking, two laser beams can differ in their alignment, i.e. in a translation δx or by an angle $\delta\theta$, they can also differ in their focussing, i.e. in waist size δw_0 or waist position δz_0, and by an astigmatic distortion, i.e. they show a difference of waist dimensions or waist positions along either the x-y axis or along the 45-degree rotated x'-y' axis denoted in the following by $\delta w_{0,\text{ord}}$ and $\delta z_{0,\text{ord}}$, or $\delta w_{0,\text{rot}}$ and $\delta z_{0,\text{rot}}$, respectively. If all these deviations are small then the normalized modal part of the complex amplitude of one of the beams can be expressed in terms of the basis of the other beam by [11]

$$\check{U}_d = U_0 \left[D_0 V_{00} + D_x V_{10} + D_y V_{01} + D_{\text{foc}} (V_{20} + V_{02}) \right. \tag{2.8}$$
$$\left. + D_{\text{ord}} (V_{20} - V_{02}) + D_{\text{rot}} V_{11} \right] \quad .$$

In this expression the contribution from relative transversal translation and tilt is described by

$$D_x = \frac{1}{w_0} (\delta x + iz_r \delta\theta_x) \quad , \tag{2.9}$$
$$D_y = \frac{1}{w_0} (\delta y + iz_r \delta\theta_y) \quad ,$$

and the contribution from a relative mismatch of the waist size or position is de-

2. High-power 532 nm single-frequency TEM$_{00}$ laser sources

scribed by

$$D_{\text{foc}} = \frac{1}{\sqrt{2}} \left(\frac{\delta w_0}{w_0} + i\frac{\delta z_0}{2z_r} \right) , \qquad (2.10)$$

$$D_{\text{ord}} = \frac{1}{\sqrt{2}} \left(\frac{\delta w_{0,\,\text{ord}}}{w_0} + i\frac{\delta z_{0,\,\text{ord}}}{2z_r} \right) ,$$

$$D_{\text{rot}} = \left(\frac{\delta w_{0,\,\text{rot}}}{w_0} + i\frac{\delta z_{0,\,\text{rot}}}{2z_r} \right) .$$

All these contributions are normalized such that the power fraction contained in the TEM$_{00}$ mode is given by

$$\begin{aligned}D_0 &= \sqrt{1 - |D_x|^2 - |D_y|^2 - 2|D_{\text{foc}}|^2 - 2|D_{\text{ord}}|^2 - |D_{\text{rot}}|^2} \\ &= \sqrt{\eta_{00}} .\end{aligned} \qquad (2.11)$$

Often these deviations are varying in time. In this case the ones connected to D_x and D_y are usually called pointing or beam jitter.

All of the above considerations are also valid for another set of transversal modes called transverse electromagnetic Laguerre Gauss (LG) modes. They make use of the Laguerre instead of the Hermite polynomials and are better suited if cylindrically symmetric problems are considered.

If transversal variations of the electric field are not important, a simpler expression for U can be used

$$U(x,y,z,t) = \overline{U}(z,t) = U_0(z,t) . \qquad (2.12)$$

The corresponding solution for E is called a 'plane wave', which is considered to have infinite lateral extension and plane phase fronts.

2.1.3. Optical resonators

An optical resonator or cavity can act as an amplifier for that part of its incident laser power, which is contained inside its eigenmode. To do so the optical phase of the eigenmode has to reproduce itself after each round trip, i.e. the resonator has to be resonant. In that case the amplification factor is given by the ratio of the

2.1. Light propagation through media

laser power inside the resonator, which is traveling into one direction P_{circ}, divided by the fraction $\eta_{00} P_{\text{inc}}$. Here $\eta_{00} P_{\text{inc}}$ is the fraction of the power P_{inc} incident on the coupling mirror or input coupler, which is contained in the resonator's eigenmode (which is assumed here to be the TEM$_{00}$ mode; see also 2.1.2). This ratio is called (resonant) power build-up

$$PB = \left. \frac{P_{\text{circ}}}{\eta_{00} P_{\text{inc}}} \right|_{\text{on resonance}} . \qquad (2.13)$$

Consider an empty linear resonator consisting of two mirrors with an optical round trip length L_{rt} fed by a laser, whose frequency difference to the resonance of the cavity is $\Delta\nu$. The coupling mirror CM has a power transmission coefficient of T_{in}. After one round trip the light power has diminished due to transmission through the end mirror T_{out}, absorption and scattering summed up into the passive fractional losses $A_{\text{p}} = T_{\text{out}} + A_{\text{p,a}} + A_{\text{p,s}}$. Another relevant loss one might think of in case of a cavity enhanced frequency doubling scheme is the loss for the fundamental light A_{SHG} caused by the effect of SHG. All these fractional power losses are combined in $A = A_{\text{p}} + A_{\text{SHG}}$.

During a round trip between the mirrors the light acquires a phase $\psi = 2\pi\nu L_{\text{rt}}/c_0$. Resonant enhancement of the circulating light power is achieved when the light wave nearly reproduces itself after one round trip, i.e. when $\Phi = \psi \bmod 2\pi \approx 0$. A cavity has therefore an infinite number of resonance frequencies spaced by the free spectral range

$$FSR = \frac{c_0}{L_{\text{rt}}} .$$

In terms of the wavelength and cavity round trip length the resonance condition is given by

$$L_{\text{rt}} = n\,\lambda_0 \quad \text{(with integer } n\text{)}. \qquad (2.14)$$

Under the assumption that T_{in}, A and Φ are all small compared to unity, the power

2. High-power 532 nm single-frequency TEM$_{00}$ laser sources

buildup can be approximated by [129]

$$PB = \frac{T_{\text{in}}}{1 + (1 - T_{\text{in}})(1 - A) - 2\sqrt{(1 - T_{\text{in}})(1 - A)}\cos(\Phi)} \quad (2.15)$$

$$\approx \frac{4\,T_{\text{in}}}{(T_{\text{in}} + A)^2 + 4\,\Phi^2} \quad ,$$

Correspondingly, the power transmitted, reflected, converted and lost is given by

$$P_{\text{trans}} = T_{\text{out}}\, PB\, \eta_{00}\, P_{\text{inc}} \quad , \quad (2.16)$$
$$P_{\text{loss}} = (A_{\text{p,a}} + A_{\text{p,s}})\, PB\, \eta_{00}\, P_{\text{inc}} \quad ,$$
$$P_{\text{h}} = A_{\text{SHG}}\, PB\, \eta_{00}\, P_{\text{inc}} \quad ,$$
$$P_{\text{refl}} = P_{\text{inc}} - P_{\text{trans}} - P_{\text{h}} - P_{\text{loss}} = P_{\text{inc}}\,(1 - A\, PB\, \eta_{00}) \quad .$$

The derivative of Eq. (2.15) in the resonant case with respect to T_{in} can be set to zero to find the maximum power build-up with respect to input coupling. This maximum is achieved if the so-called impedance matching condition is fulfilled:

$$T_{\text{in}} = A \quad \Rightarrow \quad PB_{\text{max}} = \frac{1}{T_{\text{in}}} \quad . \quad (2.17)$$

Hence, the largest power build-up is obtained when T_{in} is chosen to be as close as possible to the sum of all losses A. Derivation with respect to A shows that there is no maximum of PB over this parameter, clarifying that it is always best to keep losses as small as possible in order to maximize the power build-up. But, even when choosing the best optics available and assembling the cavity under as clean ambient conditions as possible, some passive losses A_{p} will always remain. Hence, one normally has to guess A in advance during the design process of an optical resonator to have a means to optimize PB.

To maximize the power build-up in the cavity, the alignment, shape and resonance frequency of the resonator eigenmode must be exactly matched by the incident laser beam. These parameters are defined by the alignment and radius of curvature (ROC) of the cavity mirrors and by the optical distance between them. Many single mode continuous-wave lasers emit most of their power into the fundamental transversal mode TEM$_{00}$. However, every slight mismatch in beam shape or alignment causes some fraction of the incident power to overlap with higher-order transversal eigenmodes of the cavity. In general, the higher-order spatial modes have different resonance frequencies than the fundamental mode. Hence, in such a case this power fraction is not amplified when the fundamental mode is resonant.

Small fluctuations of the cavity round trip length ΔL_{rt} induce correspondingly small relative changes of its resonance frequency given by $\Delta \nu_{res}/\nu_{res} = -\Delta L_{rt}/L_{rt}$. Thus there is a length change, which induces a difference between laser frequency and resonance frequency, at which the power build-up is reduced to half its peak value. The doubled distance between these two points on the frequency axis is denoted as full linewidth at half maximum $FWHM$ of the optical resonator and can be calculated by

$$FWHM \approx \frac{FSR}{2\pi}(T_{in} + A) \quad . \quad (2.18)$$

With this expression Eq. (2.15) can be casted in another way, which shows, that the influence of the mirrors and losses, and the influence of the frequency mismatch can be separated for small frequency deviations $\Delta \nu$:

$$PB \approx PB_{max} \cdot \frac{1}{1 + \left(\frac{\Delta \nu}{FWHM/2}\right)^2} \quad \text{with} \quad PB_{max} = \frac{4T_{in}}{(T_{in} + A)^2} \quad . \quad (2.19)$$

This frequency dependent power buildup can be utilized to filter sideband frequencies above $FWHM/2$ from the incident beam, suppressing technical amplitude and phase modulations in the circulating light.

The ratio $FSR/FWHM$, which is a measure for the cavity's quality, is usually called finesse $\mathcal{F}$.

The full theory on mode structures of laser beams and resonators can be found in several publications, for instance [129, 75, 80].

2.1.4. Propagation in a nonlinear medium

Compared to the case of linear media the situation changes, if media are considered, whose atomic response to the excitation of the transmitted electric field is to some extend nonlinear. Now, the expression for the atomic polarization might be

2. High-power 532 nm single-frequency TEM$_{00}$ laser sources

expanded into a Taylor series. It should be noted, however, that this Taylor expansion does not necessarily converge for some nonlinear processes. This is the case for resonant optical excitation (because the atomic populations are changed) and for fundamental intensities on the order of $10^{16}\,\frac{\text{W}}{\text{cm}^2}$ or bigger (because photoionization occurs). Because this thesis deals with none of these effects and is interested only in the quadratic part, the atomic polarization simplifies to its Taylor expansion [26, 39]

$$\tilde{P}_i(t) \approx \tilde{P}_i^{(1)}(t) + \tilde{P}_i^{(2)}(t) = \sum_j \epsilon_0 \chi_e^{(1)}{}_{ij} E_j(t) + \sum_{j,k} \epsilon_0 \chi_e^{(2)}{}_{ijk} E_j(t) E_k(t) \quad . \quad (2.20)$$

In this expression E_j and E_k are not necessarily polarized along the same optical crystal axis.

An intuitive understanding of the origin of the nonlinear components of the atomic polarization can be obtained by thinking of an electron as a harmonic oscillator in its ideally parabolic atomic potential (i.e. by application of the Lorentz model of atoms). Many nonlinear materials are solids where individual atoms are not free but closely packed in a grid-like structure together with atoms of different kinds. This leads to deformations of the electronic potentials, which is in the Lorentz model depicted as deformation of the parabolic potential. While a parabolic atomic potential would cause only a linear component of $\tilde{P}$ to appear, noncentrosymmetric deformations can cause components of second and higher-order. Instead, centrosymmetric distortions can cause only components of third and higher-order. Fig. 2.1 depicts a these potentials. As will be shown below the tensor structure in Eq. (2.20) can be reduced to a single scalar parameter for many crystal materials in a given experimental situation. Therefore it is ignored here for a while.

Incorporating the thoughts from the paragraph dealing with linear media, the differential equation describing the light propagation Eq. (2.2) gets a new source term on its right hand side:

$$\nabla^2 E(t) - \frac{\epsilon_r \mu_r}{c_0^2} \frac{\partial^2}{\partial t^2} E(t) = \frac{\chi_e^{(2)}}{c_0^2} \frac{\partial^2}{\partial t^2} E^2(t) \quad .$$

To understand the effect of this new source term one assumes that the electric field was prepared by the experimentalist to consist of two frequency components ν_1 and

2.1. Light propagation through media

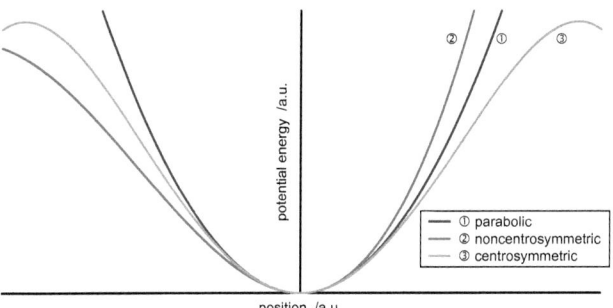

Fig. 2.1: Comparison of ideally parabolic, noncentrosymmetric and centrosymmetric atomic potentials 'seen' by an electron in the Lorentz model [26].

ν_2
$$E = \left[\frac{1}{2}U_1 e^{i(k_1 z - 2\pi\nu_1 t)} + (cc)\right] + \left[\frac{1}{2}U_2 e^{i(k_2 z - 2\pi\nu_2 t)} + (cc)\right] \quad .$$

Here the complex conjugates have to be inserted because the propagation equation of a nonlinear medium applies *nonlinear* operators to E. Because the set of functions $\exp(i2\pi\hat{\nu}t)$ forms a complete orthonormal basis one gets a separate propagation equation for each frequency component $\hat{\nu}$ of the nonlinear atomic polarization (in the following set of equations the abbreviations denote sum frequency generation (SFG) and difference frequency generation (DFG)):

SHG of $\nu_3 = 2\nu_1$: $\quad DE(\nu_3) = \dfrac{\partial^2}{\partial t^2}\left[\dfrac{\chi_e^{(2)}}{2}U_1^2\, e^{i[(2k_1)z - 2\pi\nu_3 t]} + (cc)\right] \quad ,$ (2.21)

SHG of $\nu_4 = 2\nu_2$: $\quad DE(\nu_4) = \dfrac{\partial^2}{\partial t^2}\left[\dfrac{\chi_e^{(2)}}{2}U_2^2\, e^{i[(2k_2)z - 2\pi\nu_4 t]} + (cc)\right] \quad .$

SFG of $\nu_5 = \nu_1 + \nu_2$: $\quad DE(\nu_5) = \dfrac{\partial^2}{\partial t^2}\left[\chi_e^{(2)} U_1 U_2\, e^{i[(k_1+k_2)z - 2\pi\nu_5 t]} + (cc)\right] \quad ,$

DFG of $\nu_6 = \nu_2 - \nu_1$: $\quad DE(\nu_6) = \dfrac{\partial^2}{\partial t^2}\left[\chi_e^{(2)} U_2 U_1^\dagger\, e^{i[(k_2-k_1)z - 2\pi\nu_6 t]} + (cc)\right] \quad .$

Here the general frequency dependence of $\chi_e^{(2)}(\nu)$ and the time dependence of $U(t)$ have been omitted for compactness, the wave vector is again defined as $k = 2\pi n(\nu)\nu/c_0$

2. High-power 532 nm single-frequency TEM$_{00}$ laser sources

and the equation's left hand side was defined as

$$\mathrm{DE}(\nu) = c_0{}^2 \nabla^2 E(\nu,t) - [\epsilon_r \mu_r] \frac{\partial^2}{\partial t^2} E(\nu,t) \quad .$$

In contrast to a linear medium, transmission of light through a nonlinear medium obviously causes a set of frequency components to be created.

For completeness it should be mentioned here that there is still another effect, which originates from the DC component of the nonlinear polarization

$$\tilde{P}(\nu=0,t) = \frac{1}{2}\left(|U_1(t)|^2 + |U_2(t)|^2\right) \quad .$$

This effect is called optical rectification, it causes a DC electric field to become measurable at the medium's facets [15], and it recently gained importance because it allows to generate terahertz radiation from laser light [83].

2.1.5. Nonlinear susceptibility

Often nonlinear materials are crystals and in general their second order susceptibility $\chi_e^{(2)}$ is a tensor. For a given nonlinear crystal the second order atomic polarization $\tilde{P}^{(2)}$, which is caused by a transmitted light beam, depends on the orientation of the optical crystal axes to this beam. However, in many practical cases (and in all, which are considered in this thesis) the material's resonance frequencies lie far off the optical frequencies involved [26]. As a consequence the system responds essentially instantaneously to the applied fields and the atomic polarization can be expanded into a Taylor series as was already assumed earlier. Under these circumstances the nonlinear susceptibility becomes essentially independent of the optical frequency and it is symmetric under the permutation of its indices. This situation is called Kleinman's symmetry and in this regime often the so-called contracted notation is used

$$d_{ijk} = \frac{1}{2}\chi_e^{(2)}{}_{ijk} \quad .$$

Then one can define a so-called effective nonlinearity d_{eff}. On theoretical grounds this single parameter fully characterizes the suitability of a certain nonlinear material on a microscopic scale for a given conversion process at given wavelengths and given orientation of the light propagation direction and polarization relative to the optical crystal axes. The way how d_{eff} is calculated for different kinds of materials is described for instance in [99].

2.1.6. (Quasi-)Phase matching

Up to this point all differential equations describing the propagation of waves inside a medium were only valid for an individual atom (i.e. on a microscopic scale). But any bulk crystal consists of a very large set of atoms, which all act as individual sources for tiny partial electromagnetic waves. In order to get a significant wave out of the crystal all these tiny sources (in the following called partial waves) have to interfere constructively. This is trivial in the case of a linear medium because the atomic emission is triggered by the exciting wave, which has the same frequency and thus travels at the same speed like all generated partial waves such that constructive interference is guaranteed.

However, this is not true for the case of a nonlinear medium. It was shown in connection with Eq. (2.2) that the propagation speed of a wave inside a medium is governed by the refractive index $\mathbf{n}(\nu)$, which depends on the light's frequency and polarization. Thus the primary wave at ν_1 and generated partial waves at another frequency component ν_2 will 'see' a refractive index difference $\Delta \mathbf{n} = \mathbf{n}_2 - \mathbf{n}_1$. Correspondingly, the primary wave, which drives the polarization of each atom will usually travel at a different speed through the crystal than the generated partial waves at another frequency. This means that partial waves generated at different locations along the beam path usually do not interfere constructively.

There are three ways to circumvent this problem, which are collectively denoted as phase matching:

- **Critical phase matching (CPM):** When the two wavelengths are polarized orthogonally and travel through a birefringent crystal, the refractive indices for the two wavelengths depend on the exact orientation of the polarizations relatively to the optical axes of the crystal. In some cases this can be utilized to match them for a given orientation and thus achieve critical phase matching.

The term 'critical' refers to the point that this kind of phase matching is rather sensitive to the orientation angles. This includes the divergence angles of the laser beam, which sometimes necessitates sub-optimal beam radii [106]. Furthermore in the case of critical phase matching the fundamental and harmonic beams do not travel collinearly in the medium, which limits their interaction length and tends to distort the harmonic beam shape [110, 142]. Typical examples of materials used for critically phase-matched SHG of laser light at a wavelength of 532 nm are potassium titanyl phosphate ($KTiOPO_4$) (KTP) and lithium triborate (LiB_3O_5) (LBO).

- **Noncritical phase matching (NCPM):** When each wavelength is polarized parallel to the optical axis of the crystal, the corresponding refractive index values are at their angular extrema and thus rather insensitive to angular alignment changes. For some crystal materials and wavelengths they can still be matched because they usually depend differently on the crystal temperature [106]. Typical examples of materials used for noncritically phase-matched SHG of laser light at a wavelength of 532 nm are lithium niobate ($LiNbO_3$) (LN) and again LBO.

- **Quasi-phase matching (QPM):** In the case of quasi-phase matching an entirely different concept is used. Here, the two wavelengths are usually polarized along the same optical axis of the crystal and thus see inherently different refractive indices. Consequently, they will accumulate a constantly growing phase difference. If the two waves start in phase the harmonic power will increase until the waves acquired a phase difference of π. After the corresponding propagation length (which usually is on the order of a few micrometers for wavelengths from the near-infrared (NIR) to visible) the optical axis of the crystal is inverted leading to an inverted sign of the effective nonlinearity d_{eff} [133]. Thus the phase difference will grow further but the electric fields will interfere *constructively* again, as d_{eff} is inverted (see Eq. (2.22) further down). The slope of the harmonic power will show a saddle point and starts rising again until the optical axis has to be inverted again. On the one hand this technique can lead to much stronger effective nonlinearities and a broader range of crystal materials can be accessed. On the other hand the thickness of the crystals is limited (usually to 1 mm) and parasitic processes might also be stronger [106, 26]. Typical examples of materials used for quasi-phase-matched SHG of laser light at a wavelength of 532 nm are periodically-poled KTP (PP-KTP), periodically poled LN (PPLN) and periodically poled stoichiometric lithium tantalate (LiTaO$_3$) (PPSLT).

Application of any of these techniques assures that the partial waves generated at different positions along the beam path interfere constructively again.

2.1.7. The coupled equations of motion for SHG

It was shown in connection with Eq. (2.21) that a light beam that propagates inside a nonlinear crystal can in general produce several new frequency components. But in practice the phase matching condition in common materials can often only be realized for two wavelengths. Thus only those nonlinear processes, which involve these two wavelengths will take place with significant efficiency.

So if one deliberately introduces phase matching for SHG from the fundamental frequency ν_f to the harmonic frequency ν_h then *two* processes can take place efficiently, namely the degenerate SHG/SFG process $\nu_h = \nu_f + \nu_f = 2\nu_f$ and the DFG process $\nu_f = \nu_h - \nu_f$. Thus using the set of equations around (2.21) the system of differential equations one has to consider to describe the situation completely is given by

$$\mathrm{DE}(\nu_h) = \frac{\partial^2}{\partial t^2} \left[\frac{\chi_e^{(2)}}{2} U_{0,f}^2 \mathrm{e}^{\mathrm{i}[(2k_f)z - 2\pi(2\nu_f)t]} + (cc) \right] \;,$$

$$\mathrm{DE}(\nu_f) = \frac{\partial^2}{\partial t^2} \left[\chi_e^{(2)} U_{0,h} U_{0,f}^\dagger \mathrm{e}^{\mathrm{i}[(k_h - k_f)z - 2\pi(\nu_h - \nu_f)t]} + (cc) \right] \;.$$

Modern crystal materials usually are of high quality showing only little absorption. Consequently the influence of optical absorption on the waves' amplitudes is ignored

2. High-power 532 nm single-frequency TEM$_{00}$ laser sources

here. Although its influence on the field amplitudes is negligible it can still cause significant heating of the crystal if the optical power levels are high. This effect will be considered later.

In order to get a valid model for experiments with laser beams this set of differential equations has to be solved for Gaussian beams. If one inserts Eq. (2.4), a set of differential equations for the complex amplitudes of TEM$_{00}$ beams is obtained. This involve the assumption of a continuous-wave system in its steady state (i.e. no time dependence of the complex amplitudes), TEM$_{00}$ mode beams, and the slowly varying envelope approximation (i.e. the assumption that the complex amplitudes do not change on length scales of the laser wavelength). Furthermore here the complex conjugates have been dropped on both sides. All parameters are in SI units. The equations then read:

$$\frac{\partial}{\partial z} U_{0,h}(z) = i\, K\, U_{0,f}(z)^2\, \tilde{J}(z) \quad, \tag{2.22}$$

$$\frac{\partial}{\partial z} U_{0,f}(z) = i\, K\, U_{0,h}(z) U_{0,f}(z)^*\, \tilde{J}^*(z) \quad,$$

$$K = \frac{d_{\text{eff}}\, k}{2\sqrt{\pi}\, w_{0,f}\, \mathbf{n}^2} \quad,$$

$$\tilde{J}(z) = e^{i\Delta k z} \frac{1}{1 + i\frac{z}{z_r}} \quad.$$

Here, the wave vector mismatch is defined as $\Delta k = 2k_f - k_h$. The small difference between k_f and k_h is only important for the phases but not for the amplitudes. Therefore in K the simplifications were used that $2k_f \approx k_h = k$ and $\mathbf{n}_h \mathbf{n}_f \approx \mathbf{n}^2$ with $\mathbf{n}$ taking the value of the arithmetic mean. These assumptions are valid in a broad range around the phase matching temperature.

Both, the fundamental and the harmonic mode are TEM$_{00}$ modes. However, due to the intensity dependence of the conversion process the harmonic waist is smaller than the fundamental one:

$$w_{0,h} = \frac{w_{0,f}}{\sqrt{2}} \quad. \tag{2.23}$$

2.2. Modelling high-power second harmonic generation

In this subsection two models are developed to describe a resonant SHG stage theoretically. One assumes an ideal nonlinear crystal, which is not influenced by heating from optical absorption, while the other model accounts to some extend for thermal dephasing processes, which are caused by heat deposition inside the crystal. The responsible absorption processes are explained, too.

2.2.1. The Boyd-Kleinman integral

The Boyd-Kleinman integral accounts for the intensity and phase matching of Gaussian beams inside the nonlinear crystal due to diffraction and the Gouy phase shift. For a circular TEM$_{00}$ beam with its waist at the crystal's center the integral is given by

$$\hat{h}(\sigma,\xi) = \mathrm{Im}\left(\int_{-L_k/2}^{L_k/2} \tilde{J}(z)\,\mathrm{d}z\right) = \frac{1}{4\xi}\left|\int_{-\xi}^{\xi}\frac{e^{i\sigma\tau}}{1+i\tau}\,\mathrm{d}\tau\right|^2. \qquad (2.24)$$

Here the parameters are defined as $\tau = z/z_r$, $\xi = L_k/(2z_r)$ and $\sigma = \Delta k z_r$. The parameter σ represents an arbitrary phase offset between fundamental and harmonic beam, which is in the case of an SHG stage usually chosen by the experimenter to optimize the conversion efficiency (e.g. by changing the crystal temperature). The value of $\hat{h}$ for an optimized phase matching condition $\sigma = \sigma_{\mathrm{opt}}$ was first analyzed as a function of the focussing parameter ξ by Boyd and Kleinman [25]. Their result is shown in Fig. 2.2 together with its peak value and two approximations for very large and very small ξ. The exact solution for $\hat{h}$ has a broad maximum and approximately equals ξ as long as $\xi < 0.7$, which corresponds to waists much bigger than the optimum value. The maximum value of $\hat{h} = 1.068$ is achieved for $\xi = 2.84$. This condition is often denoted as Boyd-Kleinman focussing.

As Lastzka et al. showed, the single-pass conversion efficiency can be increased above the maximum that $\hat{h}$ would allow by 4.4 %. For this to occur a temperature

2. High-power 532 nm single-frequency TEM$_{00}$ laser sources

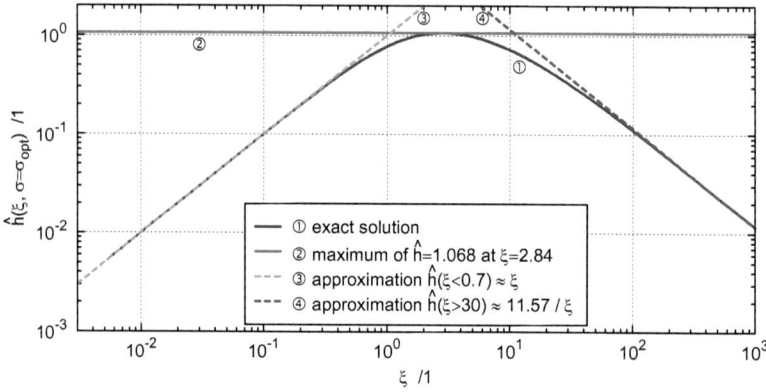

Fig. 2.2: Value of the Boyd-Kleinman integral $\hat{h}(\sigma,\xi)$ for $\sigma = \sigma_{opt}$ as a function of ξ. Also shown is its maximum value and two approximations for much larger and much smaller ξ than the optimum value.

profile has to be realized, which varies in a special way over the propagation distance inside the crystal [82].

Boyd and Kleinman also found that for small $\xi < 0.5$ and for arbitrary σ the integral can be approximated by [25]

$$\hat{h}(\sigma,\xi) = G\,\text{sinc}^2\left(\xi(\sigma - \sigma_{opt})\right) \quad, \tag{2.25}$$

with $G = \xi$ and $\sigma_{opt} = 1$. For bigger values of $\xi \leq 3.5$ there is no longer a simple expression for G but the sinc^2() dependence remains valid within the innermost 80% of its central peak and the value of σ_{opt} is approximately given by

$$\sigma_{opt} \approx \frac{\arctan(M\,\xi)}{M\,\xi} \quad \text{with} \quad M = 0.775 - 0.175\,\arctan(0.242\,\xi) \quad.$$

Under these conditions the empirically measurable temperature acceptance bandwidth δT_{bw}, which is defined as the FWHM of the central peak of the sinc^2() function under variation of the temperature of a crystal of length 1 m, can be used to substitute the argument of the sinc^2() function by

$$\xi(\sigma - \sigma_{\mathrm{opt}}) = 1.391 \frac{2\,\Delta T\,L_{\mathrm{k}}}{\delta T_{\mathrm{bw}}} \quad , \tag{2.26}$$

with the deviation ΔT of the crystal temperature from its value for optimum phase matching.

2.2.2. Approximate analytic solutions

If the generated harmonic power is small compared to the incident fundamental power, one denotes the latter as nearly 'not depleted'. In this case the coupled differential Eqns. (2.22) describing the SHG process can be solved analytically to give

$$P_{\mathrm{h}}(L_{\mathrm{k}}) = \frac{16\pi^2\,d_{\mathrm{eff}}^{\ 2}\,L_{\mathrm{k}}\,\hat{h}\,P_{\mathrm{f}}(0)^2}{\mathrm{n}^2\,\lambda_{0,\mathrm{f}}^{\ 3}\,\epsilon_0\,c_0} \quad , \tag{2.27}$$

$$P_{\mathrm{f}}(L_{\mathrm{k}}) = P_{\mathrm{f}}(0) - P_{\mathrm{h}}(L_{\mathrm{k}}) \quad .$$

These expressions for the fundamental power P_{f} and harmonic power P_{h} are well known in literature, e.g. see [26]. Here, L_{k} denotes the crystal length and $\hat{h}$ denotes the Boyd-Kleinman integral.

Spiekermann et al. proposed another relation, which was meant as substitution for Eq. (2.27), namely [134]

$$P_{\mathrm{h}}(L_{\mathrm{k}}) = P_{\mathrm{f}}(0)\,\tanh^2\left(\frac{4\pi\,d_{\mathrm{eff}}}{\mathrm{n}\,\lambda_{0,\mathrm{f}}}\sqrt{\frac{L_{\mathrm{k}}\,\hat{h}\,P_{\mathrm{f}}(0)}{\lambda_{0,\mathrm{f}}\,\epsilon_0\,c_0}}\right) \quad , \tag{2.28}$$

$$P_{\mathrm{f}}(L_{\mathrm{k}}) = P_{\mathrm{f}}(0) - P_{\mathrm{h}}(L_{\mathrm{k}}) \quad .$$

The origin of this equation is not clear. It certainly cannot be correct in any case, as it predicts P_{h} to be strictly monotonic over L_{k}, which is not the case for the coupled set of differential Eqns. (2.22). Computational tests showed that it gives acceptably precise results roughly for $\xi < 3$ and $(P_{\mathrm{f}}(0)\,d_{\mathrm{eff}}) < 1\times 10^{-9}\,\frac{\mathrm{Wm}}{\mathrm{V}}$. In this range of values it incorporates depletion of the fundamental beam and is thus much more precise than Eq. (2.27).

2. High-power 532 nm single-frequency TEM$_{00}$ laser sources

2.2.3. Doubling schemes and suitable crystals

A second harmonic generator can be designed in three distinct ways which are usually referred to as *single-/multi-pass*, *intracavity* and *external-cavity* doubling. In the case of *single-/multi-pass* doubling the fundamental laser light is simply directed once or several times through a nonlinear crystal placed downstream of the laser. The advantage of this technique is its simplicity. On the other hand for obtainable nonlinear crystal lengths of typically several centimeters the material must show a very high nonlinearity to convert a large fraction of the incident fundamental light. There are nonlinear materials like lithium niobate (LiNbO$_3$) (LN) or potassium titanyl phosphate (KTiOPO$_4$) (KTP) that show such high effective nonlinearities of $d_{\text{eff}} \approx 4-5\,\frac{\text{pm}}{\text{V}}$ in bulk [154, 134]. By periodical poling still bigger nonlinearities can be achieved, reaching values of $d_{\text{eff}} \approx 10\,\frac{\text{pm}}{\text{V}}$ for periodically-poled KTP (PPKTP) (see [134]) and periodically poled stoichiometric lithium tantalate (LiTaO$_3$) (PPSLT) (see [90]) and even $d_{\text{eff}} \approx 16\,\frac{\text{pm}}{\text{V}}$ for periodically poled LN (PPLN) [90].

Another promising way to set up an SHG is to place a crystal inside an optical resonator with a sufficiently high power buildup. The optical resonator recirculates the fundamental light not converted in previous passes through the crystal and thus enhances its input power by coherent superposition (see 2.1.3). This effect allows to use also materials with low nonlinearity, which in turn might have a higher tolerance to optical power, because the power buildup compensates for the smaller effective nonlinear coefficient (see Eq. (2.27)). Examples for such materials with lower nonlinearity that are readily obtainable are lithium triborate (LiB$_3$O$_5$) (LBO) with $d_{\text{eff}} \approx 1\,\frac{\text{pm}}{\text{V}}$ [104], beta barium borate (β-BaB$_2$O$_4$) (BBO) with $d_{\text{eff}} \approx 2\,\frac{\text{pm}}{\text{V}}$ [103] and bismuth triborate (BiB$_3$O$_6$) (BiBO) with $d_{\text{eff}} \approx 3\,\frac{\text{pm}}{\text{V}}$ [43].

This design may be realized by stabilizing the primary laser light to an additional resonator, which comprises the nonlinear crystal, which is usually denoted as *external-cavity* doubling [153]. It prevents unwanted couplings between SHG and laser process but needs some control loop for the stabilization.

Finally, the optical resonator can also be realized by the laser resonator itself as it is the case in *intracavity* doubling [56]. This scheme has the advantage that no additional control loop and optical resonator is needed. But it suffers from the so-called *green problem*, which generally causes the laser to become instable [12]. The green problem denotes a coupling between various longitudinal modes in the laser resonator due to the nonlinearity of the crystal, which causes large power fluctuations. Solution to this requires either to tolerate the emission of a large number of longitudinal modes (which reduces the effect by averaging over many contributions) or to add several additional optical components to the laser resonator to force strict single-mode emission. Such additional components tend to degrade the laser efficiency, scalability of the output power and the emitted beam shape due to effects that are caused by absorption of optical power.

2.2.4. Processes that lead to thermal dephasing

There are many examples reported in literature, which demonstrate, that the performance of SHG stages often does not follow the ideal theoretical prediction by Eq. (2.22) [56, 145, 134, 90, 132, 140, 78, 71, 120]. Many authors relate the deviations to excessive heat deposition inside the nonlinear crystal.

Deposition of heat

The most important absorption mechanisms, which cause the deposition of heat inside the nonlinear crystal of an SHG stage, are listed in the following.

Passive absorption. Although the optical quality of most nonlinear crystals has reached very high levels, the linear absorption coefficients do certainly not vanish completely. Especially those materials with high effective nonlinearity like (PP)KTP, PPSLT or (PP)LN also show significant absorption at wavelengths below and in the visible spectral range.

For a laser wavelength of 532 nm values reported in literature are on the order of $\alpha_{l,h} = 0.5 - 4.0 \frac{\%}{cm}$ [84, 92, 78]. For materials with lower nonlinearity like LBO this value can be as low as $0.005 \frac{\%}{cm}$ [92].

2. High-power 532 nm single-frequency TEM$_{00}$ laser sources

The linear absorption coefficients at 1064 nm are generally rather small. Reported values are in the range $\alpha_{l,f} = 3\times 10^{-3} - 0.3\,\frac{\%}{\text{cm}}$ [92, 136, 78].

Gray tracking. The term gray tracking denotes the creation of defect sites (also known as color centers) in the crystal lattice, which are related to the formation of excitons, i.e. electron-hole pairs [30]. In KTP, LN PPSLT and other materials these defects give rise to broad overlapping absorption bands around 510 nm and 420 nm in the green-blue visible spectral region and thus represent a photochromic damage [145]. The defects are created by two photon absorption processes initiated in turn by precursor defects in the as-grown crystal material [84]. Higher quality crystals should accordingly be more resistant against gray tracking [101].

Once initiated, this effect is expected to be roughly described by the rate equation [84]

$$\frac{\partial N}{\partial t} = \beta \left(N_s - N \right) I_{w,h}^2 \quad ,$$

with the defect creation coefficient β, the steady-state defect density N_s and the squared harmonic intensity at the crystal center $I_{w,h}$. This means that the density of defects starts at an unknown level given by the precursor defects and tends to reach N_s on long time scales. This rate equation subsequently gives rise to a time and intensity dependent nonlinear absorption coefficient given by

$$\alpha_h = \alpha_{l,h} + \alpha_{gt,h} \left(1 - e^{-\beta I_{w,h}^2 t} \right) \quad . \qquad (2.29)$$

In the case of KTP it was shown, that this effect can be cured by annealing [101]. In contrast, in PPKTP the effect appeared to be unrecoverable. For this material it was found to become important above harmonic peak intensities of only 70 $\frac{\text{kW}}{\text{cm}^2}$ [84]. However, below this intensity it seemed to be not existent. Thus gray tracking might be avoided if the harmonic intensity is kept below this value!

As stoichiometric materials like (PP)stoichiometric lithium tantalate (LiTaO$_3$) (SLT) generally have lower defect density they have higher resistance to gray tracking [132]. A similar effect could be achieved for some materials by doping with magnesium oxide.

2.2. Modelling high-power second harmonic generation

Finally, no reports are known to the author, which state that gray tracking might be an issue in LBO for any reasonable intensity.

Green-induced infrared absorption. The term green-induced infrared absorption (GRIIRA) denotes a dependence of the absorption coefficient for infrared light on the harmonic intensity present at the same place as the absorbed fundamental beam

$$\alpha_f = \alpha_{l,f} + \alpha_{griira,f} I_{w,h} \quad . \tag{2.30}$$

This effect shows up in many nonlinear materials and is expected to be at least partly related to the same defect sites, which cause gray tracking, too [145]. Correspondingly, it is also very susceptible to the crystal quality and stoichiometry and could also be reduced in some cases by magnesium oxide doping [51, 64].

While GRIIRA is an important effect in undoped (PP)LN even at rather low harmonic intensities [51], Kumar et al. found that it was of no importance in PPKTP and PPSLT for harmonic intensities on the order of $0.3 \frac{MW}{cm^2}$ [78].

Again, no reports are known to the author, which state that GRIIRA might be an issue in LBO for any reasonable intensity.

Effects of heat deposition

In the following the detrimental optical effects are listed, which are caused by the heat deposition inside the nonlinear crystal and which subsequently limit the performance of SHG experiments.

Photorefractive damage. The effect of photorefractive damage (also known as optical damage) causes a distortion of the lateral beam shape during transmission through the crystal [144]. The distortion is caused by temperature dependent refractive index inhomogeneities inside the crystal material. The thermal gradients in turn are caused by absorption from a TEM_{00} beam with Gaussian intensity distribution. During transmission the beam's wavefront undergoes a space-dependent delay, which can on the one hand lead to astigmatism and ellipticity of the beam. On the other hand it can imprint a space dependent polarization state on the beam, which is converted into a distorted intensity profile by any polarization dependent process (like SHG) exhibited afterwards [149]. This effect is more or less observable in any material, which is traversed and thus heated by a nonuniform intensity distribution.

In the field of SHG congruent undoped lithium niobate and tantalate are especially prone to show this degradation effect, which can be cured, however, by doping [73].

Thermal dephasing. Heating of the nonlinear crystal results in changes of the optical path length by temperature dependent refractive index changes and stress-induced birefringence [94]. These effects will generally be different for each interacting wavelength due to the dispersion of the nonlinear materials. Hence they will deteriorate the phase matching. The experimenter can usually compensate for constant mismatches by optimization of the overall crystal temperature. However, due to absorption from a Gaussian intensity profile, a spatially varying phase mismatch will be created, which can not be compensated for easily. In the following subsection a model is developed, which accounts to some extend for thermal dephasing.

2.2.5. Models with and without thermal dephasing

As explained previously the optical intensity inside the crystal should not be higher than necessary because it is a source of thermal distortions and efficiency degradations.

In external-cavity doubling schemes the fundamental light, which is not converted by the crystal in a single round trip is 'recycled' by the optical resonator and thus not lost for the conversion process. As long as the passive round trip losses of the SHG resonator for the fundamental beam are kept much smaller than the losses due to conversion, the intensity inside the crystal can be reduced without significant reduction of the external conversion efficiency. This will be explained in the following paragraphs.

In intracavity doubling schemes this possibility is usually limited. In addition to the nonlinear crystal such cavities contain at least the laser medium and often more optical components, which tend to increase their passive round trip losses.

2.2. Modelling high-power second harmonic generation

Additionally, in this case the choice of beam radii is constrained by the coupling of the SHG process to the laser process.

In single-/multi-pass doubling the intensity of fundamental and harmonic light inside the crystal is directly linked to the conversion efficiency of the overall device because there is no resonator.

Thus the scheme of external-cavity doubling offers an additional degree of freedom to reduce the impact of the intensity dependent detrimental effects listed in 2.2.4.

Ideal external-cavity doubling

Small depletion of fundamental. Let's gain an intuitive understanding of this possibility to reduce the intensity inside the crystal without drawbacks. Let us assume that the fundamental light is hardly depleted and thus the conversion process in the crystal (which is located now inside an external cavity) can be described by Eq. (2.27). To incorporate the effect of the surrounding optical resonator into these equations, which is assumed to be resonant with the incident light, the following substitution is made

$$P_f = PB_{max}\,\eta_{00}\,P_{inc} \quad,$$

which leads to the starting expression for the fractional round trip losses due to conversion inside the cavity of

$$\frac{P_h}{T_{dc}P_f} = A_{SHG} = M\,\frac{n}{\lambda_{0,f}}\,L_k\,\hat{h}\,PB_{max}\,\eta_{00}\,P_{inc} \quad. \tag{2.31}$$

As usual subscripts of 'f' denote parameters of the fundamental light, while such of 'h' denote harmonic light parameters. Here T_{dc} is the harmonic power transmission coefficient of the dichroic mirror of the SHG resonator, which separates the harmonic beam from the circulating infrared beam. Correspondingly P_h denotes the power outside of the cavity. The factor M is given by

$$M = \frac{16\pi^2\,d_{eff}^2}{n^3\,\lambda_{0,f}^2\,\epsilon_0\,c_0} \quad.$$

As one is interested in lowering the intensity by enlarging the waist of the circulating fundamental light, the conversion will not be done with Boyd-Kleinman focussing

2. High-power 532 nm single-frequency TEM$_{00}$ laser sources

(see 2.2.1), and therefore the approximation of $\hat{h} \approx \xi = L_k/(2z_r)$ is applied, which was shown in Fig. 2.2.

Furthermore the optical resonator is assumed to be operated in its impedance matched state. As was found in 2.1.3, the power buildup of an optical resonator is maximized in this state. This should be desirable for the working point of any resonant SHG stage as it obviously maximizes the harmonic power, too. According to this the following substitution is made

$$PB_{\max} = \frac{1}{A_{\text{SHG}} + A_p} \quad .$$

If all these considerations are included into Eq. (2.31) one arrives at the following expression for the fractional round trip losses due to conversion

$$A_{\text{SHG}} = \frac{1}{2}\left(\sqrt{A_p^2 + \frac{2}{\pi w_{0,f}^2} M L_k^2 \eta_{00} P_{\text{inc}}} - A_p\right) \quad . \tag{2.32}$$

The effect of SHG is nonlinear and correspondingly its efficiency ideally grows with increasing fundamental power. Therefore the latter equation depends on P_{inc}. In high-power resonant SHG stages the fractional round trip losses due to conversion are therefore often rather big and the approximation

$$\frac{L_k}{w_{0,f}}\sqrt{\frac{1}{2\pi} M \eta_{00} P_{\text{inc}}} =: \widetilde{A}_{\text{SHG}} \gg A_p \tag{2.33}$$

is valid, especially if high-quality optics with low passive losses are applied. With this approximation the following expressions are obtained for the key parameters of a resonant SHG stage with external-cavity doubling scheme. These are the fractional round trip losses due to conversion A_{SHG}, the circulating fundamental peak intensity in the waist at the center of the nonlinear crystal $I_{w,f}$, the external conversion efficiency observed from outside of the SHG resonator η_{SHG}, and the circulating

2.2. Modelling high-power second harmonic generation

harmonic peak intensity in the waist at the center of the nonlinear crystal $I_{w,h}$:

$$A_{\text{SHG}} \approx \widetilde{A_{\text{SHG}}} \quad , \qquad (2.34)$$

$$I_{w,f} = PB_{\max} \frac{2\eta_{00} P_{\text{inc}}}{\pi w_{0,f}^2} \approx \frac{2}{w_{0,f} L_k} \sqrt{\frac{2\eta_{00} P_{\text{inc}}}{\pi M}} \quad ,$$

$$\eta_{\text{SHG}} = \frac{P_h}{P_{\text{inc}}} \approx T_{\text{dc}} \eta_{00} \quad ,$$

$$I_{w,h} = \frac{2 P_h}{\pi T_{\text{dc}} w_{0,h}^2} \approx \frac{4 \eta_{00} P_{\text{inc}}}{\pi w_{0,f}^2} \quad .$$

In the last expression Eq. (2.23) was used.

Eq. (2.34) clearly shows that, as long as the passive losses A_p are much smaller than the approximate conversion losses $\widetilde{A_{\text{SHG}}}$, both intensities decrease with growing fundamental waist size while the external conversion efficiency will stay approximately constant! The fundamental intensity will fall as $1/w_{0,f}$ for waist sizes above the Boyd-Kleinman focussing condition, and the harmonic intensity will even fall as $1/w_{0,f}^2$. Also the fractional conversion losses A_{SHG} fall as $1/w_{0,f}$ and will come closer to the passive fractional losses. This will render the just obtained results invalid above a certain waist size.

Making the crystal longer will also have a comparably beneficial effect on the fundamental intensity but not on the harmonic intensity. It will also further increase the conversion losses. Finally, it is interesting to note, that the circulating intensity of an SHG resonator rises only with the square root of the incident power.

To obtain a means to estimate the lowest order influence of passive round trip losses and increased waist size on the external conversion efficiency one has to relax the condition of Eq. (2.33) to

$$\widetilde{A_{\text{SHG}}}^2 \gg \frac{A_p^2}{4} \quad .$$

2. High-power 532 nm single-frequency TEM$_{00}$ laser sources

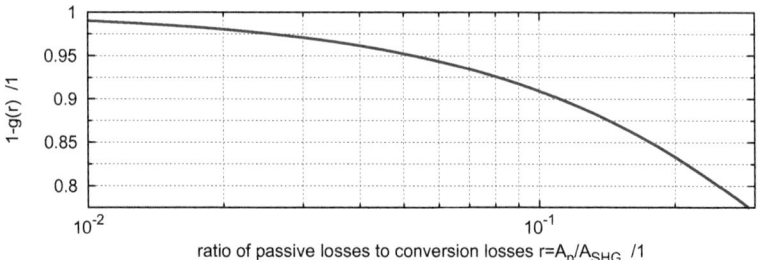

Fig. 2.3: Scaling of external conversion efficiency with ratio of passive losses to conversion losses due to Fig. 2.35.

The resulting expressions are

$$A_{\text{SHG}} \approx \widetilde{A_{\text{SHG}}} \left(1 - \frac{A_{\text{p}}}{2\widetilde{A_{\text{SHG}}}}\right) \quad , \tag{2.35}$$

$$\eta_{\text{SHG}} \approx T_{\text{dc}}\, \eta_{00} \left[1 - g\left(\frac{A_{\text{p}}}{A_{\text{SHG}}}\right)\right] \quad ,$$

$$g(r) = \frac{r}{1+r} \quad ,$$

and the function $g(r)$ is plotted above the ratio $r = A_{\text{p}}/A_{\text{SHG}}$ in Fig. 2.3. For a realistic SHG resonator with depletion of the fundamental the dependence of η_{SHG} on r will be flatter. Summarizing, the external conversion efficiency is limited by three parameters, namely by r, by the harmonic transmission coefficient T_{dc} of the dichroic mirror, which splits the two frequency components, and by the fraction η_{00} of the incident power, which is contained in the TEM$_{00}$ mode.

Including depletion of fundamental. For a more precise model of external-cavity doubling one has to incorporate depletion of the fundamental wave. As this cannot be done analytically anymore one now has to start with the set of coupled differential Eqns. (2.22). The fractional round trip losses due to conversion, which were previously obtained via Eq. (2.32) have to be obtained numerically now.

2.2. Modelling high-power second harmonic generation

To achieve this it was started from a guess of the value of A_{SHG}. As the passive losses and input coupler transmission were assumed to be known, a value for PB_{max} could be calculated via Eq. (2.15). From these values the harmonic power obtained from the circulating power in a single pass through the crystal was derived via the coupled differential equations. Finally, a refined value for A_{SHG} could be obtained from the ratio of harmonic power and circulating power. This procedure was then repeated until the power buildup values effectively did not change anymore. From this numerical result the other key parameters of a resonant SHG stage with external-cavity doubling scheme, which were also derived in the previous paragraph, could be obtained.

Correspondingly, this model allows the simulation of a resonant SHG stage with external-cavity doubling design. It incorporates the interaction of Gaussian beams, possible depletion of the fundamental beam and passive losses inside the cavity. However, it is assumed that the single-pass efficiency of the nonlinear crystal is not influenced by heating from optical absorption.

Including thermal dephasing at high intensities

In 2.2.4 several processes were explained, which are caused by heating from optical absorption inside the crystal, and which lead to a spatially varying phase mismatch between fundamental and harmonic light. This is denoted as thermal dephasing and causes a reduction of the harmonic output power of an SHG stage. Here, a basic extension of the above described model is developed, which incorporates thermal dephasing effects to a certain extend.

The following model extension is based on a single-pass SHG model by Liao et al., who proposed to assign an effective average value $\overline{\Delta\sigma}$ to the transversally varying phase mismatch and then to apply Eq. (2.25) with $\sigma = \sigma_{\text{opt}} + \overline{\Delta\sigma}$ to model its deteriorating influence on the harmonic output power of a single-pass SHG stage under weak focussing conditions [84].

This model was adopted here for the case of resonant SHG. First the temperature gradient in the crystal was calculated and from this $\overline{\Delta\sigma}$ was obtained via Eq. (2.26).

2. High-power 532 nm single-frequency TEM$_{00}$ laser sources

To enhance the validity of the original model to stronger focussing, not Eq. (2.25) was used, but the detuned value of σ was directly inserted into the coupled differential Eqns. (2.22). They were subsequently used to calculate the harmonic output power of the resonant SHG stage as described above for the model without thermal dephasing.

Assuming that the absorption coefficients for fundamental and harmonic light are known, the algorithm was then complemented by an additional outer iteration loop. Whenever the inner loop had succeeded in finding a pair of values for A_{SHG} and $PB_{\max}$ for an initial guess of the phase matching detuning $\overline{\Delta\sigma}$, these values were used to calculate the correspondingly absorbed power inside the crystal, and from that a refined detuning $\overline{\Delta\sigma}$ was derived. The outer loop was repeated until the complete parameter set did no longer change. The set of equations, which was used to do these outer loop calculations was the following:

$$\overline{\Delta\sigma} = 5.568\, z_r\, \frac{\Delta T}{\delta T_{\text{bw}}} \quad , \qquad (2.36)$$

$$\Delta T = \frac{P_{\text{heat}}}{4\pi\, L_k\, \kappa_{\text{th}}}\, [0.577 + \ln(4)] \quad ,$$

$$P_{\text{heat}} = \left(1 - e^{-\alpha_h L_k}\right) P_h^{(\text{generated})} + \left(1 - e^{-\alpha_f L_k}\right) P_{\text{circ}} \quad ,$$

$$P_h^{(\text{emitted})} = e^{-\alpha_h L_k}\, P_h^{(\text{generated})} \quad .$$

Here, δT_{bw} denotes the thermal acceptance bandwidth of SHG of light with 532 nm wavelength (5.94×10^{-2} Km for PPKTP [134, 78]), κ_{th} denotes the thermal conductivity of the crystal material (3 $\frac{\text{W}}{\text{Km}}$ for PPKTP [71]), d_h the height of the crystal, and P_{heat} denotes the heating power.

Heating from the fundamental as well as from the harmonic beam was considered as source of thermal dephasing. Although the fundamental absorption coefficient is generally very small, its contribution might still be important as the circulating power is usually quite high and as the effect of thermal dephasing is nonlinear such that even small contributions above a certain threshold can have significant effects. The expressions for the absorption coefficients are given by Eq. (2.30) for α_f and by Eq. (2.29) for α_h. Their numeric values are either known or were determined by fitting the model to experimental data measured within the work of this thesis.

The spatial heat source profile was assumed to be dominated by the dimensions of the harmonic beam as its absorbed power was still expected to be larger than the fundamental one for the experiments done in the scope of this thesis. As phase matching between harmonic and fundamental beam has to be achieved and as the fundamental beam radius is bigger, the overall thermal gradient ΔT was taken as the temperature difference between the beam center and the fundamental waist, which emerges into the bracketed term in the ΔT equation. The general expression for this thermal gradient was derived from [94, 77, 84] to the leading order. It assumed a rod of material, whose radius is much bigger than the heat source.

Finally, P_h is corrected for the absorption, which created the thermal gradient. Due to the very small absorption coefficients of modern nonlinear crystals for 1064nm light, the circulating fundamental power was not corrected for this absorption.

The model does not account for the spatial variation of the harmonic power along the propagation distance inside the crystal. Furthermore, it does not account for the effect of stress-induced birefringence, which will also originate from any thermal gradient inside the crystal [93].

2.3. Converting an intermediate power metrology laser

In 2.2.5 it was explained that external-cavity doubling schemes offer an additional degree of freedom compared to single-pass SHG stages, which allows to reduce the fundamental and harmonic intensities inside the nonlinear crystal without sacrificing the external conversion efficiency. This concept was tested by conversion of an intermediate power single-frequency metrology laser, which provided a maximum power of 10.2 W at a wavelength of 1064 nm. More than 90 % of this power were emitted into the fundamental transversal mode.

As nonlinear material for this test PPKTP was chosen. On the one hand it was obtainable with lengths of up to 2cm and it combines a high nonlinearity, a low linear absorption coefficient for 1064nm light and quasi-phase matching. On the other hand

2. High-power 532 nm single-frequency TEM$_{00}$ laser sources

several authors had already published results, which hinted on intensity dependent detrimental effects inside such crystals like gray tracking (e.g. see [84, 30, 101]) or GRIIRA (e.g. see [145, 84, 134]) or thermal dephasing in general [84, 94] (for all these effects see also 2.2.4). Thus it was interesting to test its performance when the fundamental and harmonic intensities were lowered by more than an order of magnitude by utilization of an external-cavity doubling scheme with enlarged fundamental waist size inside the crystal.

2.3.1. Reported green light power levels from PPKTP

To date several publications on SHG of continuous-wave single-frequency laser light at a wavelength of 532 nm with PPKTP crystals are available, for example [125, 77, 56]. The highest achieved harmonic powers known to the author are 6.2 W from single-pass experiments [125, 78], and 2.3 W from intracavity experiments [56]. The high single-pass power reported in [125, 78] degraded by about 20 % within the first 60 minutes of operation and is assumed to have suffered further degradation over longer timescales.

High conversion efficiencies above 40 % were only obtained from external-cavity experiments, but their harmonic output powers remained below 1 W as was for instance reported in [71].

2.3.2. General external-cavity design considerations

Experimental constraints

The resonator for a second harmonic generator with the desired high output power had to be designed as a compromise between several constraints. First, thermal gradients inside the nonlinear crystal had to be kept small to avoid degradation of SHG efficiency (as explained in [84, 90]) and the laser-induced destruction of the crystal or of dielectric coatings had to be avoided. The mirrors and crystals used for the SHG experiments presented in this thesis were coated by the electron beam deposition coating technique [53]. Their laser-induced damage thresholds for CW laser light depended on the actual manufacturer and optic used and were stated to be on the order of approximately $2-5\,\frac{\text{MW}}{\text{cm}^2}$.

Second, the fundamental waist inside the crystal had to be designed such that it shows no astigmatism to avoid degradation of the conversion efficiency (as was already found by Boyd and Kleinman [25]) and to avoid ellipticity of the harmonic beam. For the same reasons only noncritically or quasi-phase-matched crystals should be used because a walk-off angle between fundamental and harmonic beam tends to distort the beam shape [142].

Third, some more practical constraints arose from the commercially available crystal lengths and radii of curvature of the resonator mirrors, the dimensions of the crystal oven in combination with a desirable compact resonator design, the maximum tolerable angle of incidence on the resonator mirrors, which are usually coated for normal incidence, and, of course, a desirable high conversion efficiency.

The SHG resonators described in this thesis were always bow-tie resonators, which usually are more flexible in achieving the desired waist size than linear resonators. Moreover, a linear resonator has the disadvantage that the harmonic light has to travel twice through the crystal in order to avoid the generation of two equally strong harmonic beams. In this case one end facet of the crystal has to be coated for high reflectivity at fundamental and harmonic wavelength, because otherwise the dispersion of air might degrade the phase matching. Apart from the additional efforts for the crystal manufacturing, this coating generally introduces a phase shift between fundamental and harmonic wave, which complicates the phase matching condition and thus the modelling of the SHG stage [82]. Additionally in such a case the fundamental intensity seen by the crystal is higher by a factor of two and also the harmonic intensity is higher.

Optimization procedure

The solution to these constraints was found as described in the following. One started from an initial design with low circulating intensity. That was obtained by two steps. First, the single-pass conversion efficiency approximately given by

2. High-power 532 nm single-frequency TEM$_{00}$ laser sources

Eq. (2.27) was optimized by assuming the use of the longest commercially obtainable crystal. Second, the SHG resonator was simulated as described in 2.2.5 to find a range of fundamental waist sizes inside the nonlinear crystal, which significantly reduce fundamental and harmonic intensities and simultaneously have little impact on the external conversion efficiency. As the absorption coefficient of the crystals at 532 nm was not known, here the model was used, which assumed no heating of the crystal.

With the knowledge of crystal length and suitable range of fundamental waist sizes the final design was found by iteratively calculating possible resonator designs by standard ABCD matrix formalism (e.g. see [129, 75]) and testing, if they fulfilled the experimental constraints listed above, until a satisfactory solution was found.

2.3.3. Design of the resonant PPKTP SHG stage

Most of the design considerations necessary were already listed in the previous subsection. The chosen nonlinear material PPKTP is quasi-phase-matched with fundamental and harmonic wave polarized along the crystal's optical z-axis. This implies that the refractive indices are not matched. For a phase matching temperature of 38 °C the values are $n_f = 1.83$ and $n_h = 1.89$ [92]. As was described around Eq. (2.22) it was sufficiently precise here to work with a single average value of $n \approx 1.86$. The crystal's value of d_{eff} for fundamental light polarized parallel to its optical z-axis was measured in a single-pass experiment to be $d_{eff} = 8.4 \frac{pm}{V}$ at optimum phase matching temperature of 38 °C (see Fig. 2.4).

Together with the four mirrors of the bow-tie resonator, six facets were 'seen' by the light at each round trip, which were all polished to a planarity of $\lambda/4$ according to the international standard MIL-O-1380A and were afterwards coated by the electron beam deposition coating technique [53]. From experiments with comparable mirrors at another wavelength their passive loss due to scattering and absorption of $A_{p,s} + A_{p,a} = 0.14\%$ was known. Thus an approximate passive roundtrip loss of $A_p = 1\%$ was assumed. Tbl. 2.1 summarizes all parameters and their values, which were used in the model described in 2.2.5 to simulate this resonant SHG stage. The simulation

2.3. Converting an intermediate power metrology laser

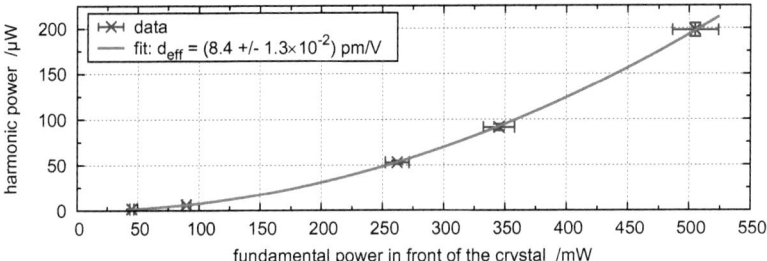

Fig. 2.4: Measurement of the harmonic power obtained from a single-pass of a fundamental laser beam with a waist of $w_{0,f} = 225$ µm at the center of the PPKTP crystal used for resonant SHG. The crystal's effective nonlinearity d_eff was determined from a fit of Eq. (2.27) to the data.

results are shown in Fig. 2.5. In this simulation the transmission of the input coupler was set to fulfill impedance matching of the SHG resonator for each individual waist size. This is meaningful because in a real experiment the experimenter would choose the input coupler transmission such that the cavity is impedance matched and the harmonic power maximized after he has decided on the actual waist size he wants to use.

In each of the three graphs belonging to Fig. 2.5 two vertical lines are shown, which mark the fundamental waist size corresponding to Boyd-Kleinman focussing (blue) and that one, which was chosen for the experiment described below (dark yellow). One can clearly see from the topmost graph that the fractional round trip losses due to conversion are highest when Boyd-Kleinman focussing is chosen. It is also obvious that they stay well above the assumed passive fractional round trip losses of $A_p = 1\%$ in a very broad region of waist sizes. This is confirmed by the second graph, which shows that the expected harmonic power shows only small variations in the same region. However, the third graph shows that harmonic and fundamental intensities fall rapidly within the same region, making it advantageous to choose a considerably larger fundamental waist size than what is corresponding to

2. High-power 532 nm single-frequency TEM$_{00}$ laser sources

parameter	value
L_k	2.0 cm
n	1.86
$\lambda_{0,f}$	1064 nm
P_{inc}	10.2 W
T_{dc}	95 % at 532 nm
A_p	1.0 %
d_{eff}	8.4 $\frac{pm}{V}$
η_{00}	95 %
T_{in}	imp. matched

Table 2.1: List of measured as well as estimated design values for experimental parameters relevant for the simulation of the resonant PPKTP SHG stage. The values of P_{inc}, A_p and T_{in} are only estimations here. Their measured values are given further below.

Boyd-Kleinman focussing. For the experiment described in the following the waist size was increased by a factor of nine (i.e. to 220 µm) compared to Boyd-Kleinman focussing (i.e. to 25 µm), achieving a reduction in the fundamental intensity by a factor of 18 and a reduction in the harmonic intensity by even a factor of 90. The corresponding expected reduction in the harmonic power is only 13 %. A further increase of the fundamental waist was not possible as the clear height of the PPKTP crystal was only about 0.9 mm. One should note here that for this high-power doubling experiment only the choice of such a big fundamental waist could reduce the harmonic intensity below the threshold values for GRIIRA and gray tracking known from literature and given in 2.2.4!

The design value for the power transmission of the input coupler was also obtained from this figure as the sum of the conversion losses for a waist of 220 µm plus the assumed passive losses of A_p.

2.3. Converting an intermediate power metrology laser

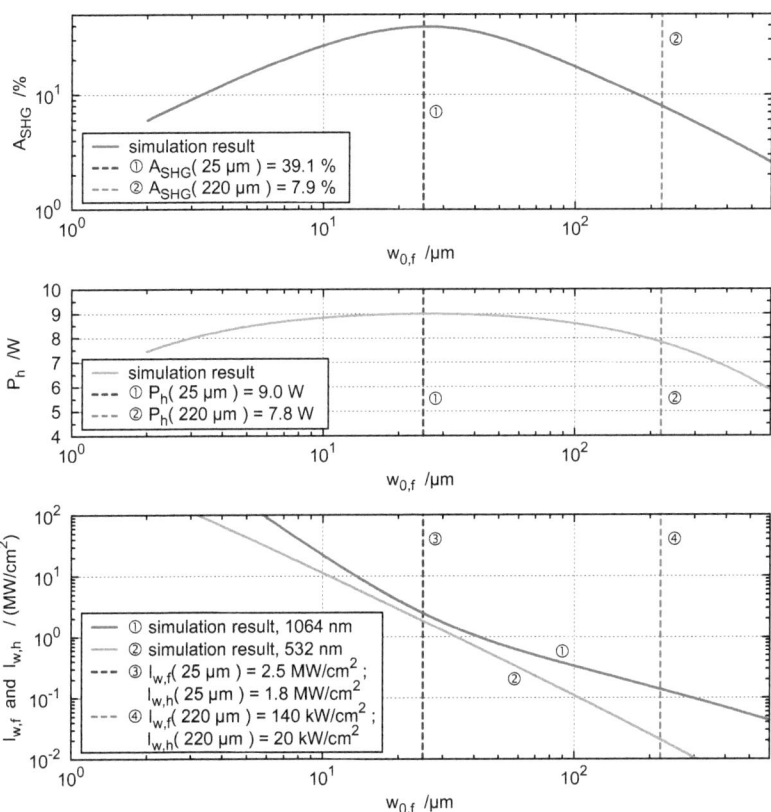

Fig. 2.5: Simulation of the PPKTP external-cavity SHG stage neglecting thermal dephasing (see 2.2.5 and Tbl. 2.1 for model and parameters). Expected conversion losses (top), harmonic power (middle) and peak intensities (bottom) are plotted depending on the fundamental waist size. The vertical lines depict the waist size corresponding to Boyd-Kleinman focussing (blue, (1)) and the one chosen for this experiment (yellow, (2)).

2.3.4. Experimental setup

The optical setup of the experiment is shown in Fig. 2.6. The infrared fundamental light from the intermediate power metrology laser first passed $\lambda/2$ waveplate P1 and polarizing beamsplitter Pol-BS1, which formed a variable attenuator to set the fundamental power incident on the SHG resonator (the beam dump path is omitted for clarity). A constant fraction of this incident power was sensed by photodetector PD1. The incident polarization state was also cleaned by Pol-BS1 and afterwards set to maximum conversion efficiency with $\lambda/2$ waveplate P2. The lenses L1 and L2 were used together with alignment mirrors (not shown) to match the incident mode to the eigenmode of the optical resonator. Photodetector PD2 measured the light, which was reflected from the SHG resonator, photodetector PD3 measured the light circulating inside it, and photodetector PD4 measured the harmonic power available outside of the SHG resonator.

The plano-concave resonator itself consisted of four one inch mirrors in a bow-tie configuration comprising the nonlinear crystal inside its oven. Three of the resonator

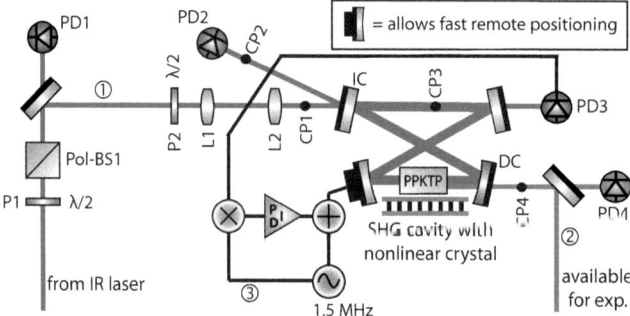

Fig. 2.6: Schematic overview of optical setup of PPKTP based resonant SHG stage. Red lines (1) depict fundamental infrared light, green lines (2) depict harmonic visible light, and black lines (3) denote electric wiring. See the text for a description of the components.

2.3. Converting an intermediate power metrology laser

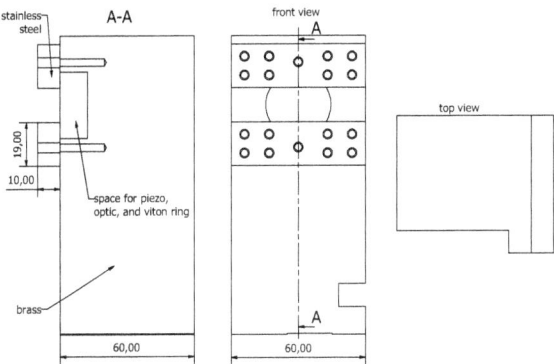

Fig. 2.7: Design drawing of the self-made rigid mount for piezo-electric transducer and one inch mirror. All given length values are in units of millimeters. Its first relevant mechanic resonance with mirror and piezo installed was above 20 kHz.

mirrors were mounted in high quality commercial mirror mounts. The fourth mirror was mounted on top of a piezo-electric transducer (also simply called piezo), which allowed for fast position changes to match the resonance frequency of the cavity to the frequency of the incident laser light. The combination of piezo and mirror was clamped together with a viton ring into a very rigid mount. The mount was optimized to show its lowest frequency relevant mechanical resonance above 20 kHz when completely assembled. A design drawing of the corresponding mount is shown in Fig. 2.7. By this technique three of the four mirrors remained adjustable and replaceable. The first point eased the alignment procedure, in which an eigenmode had to be established without misaligning the beam inside the nonlinear crystal. The second point was important for the input coupler, which had to be exchanged to realize an impedance matched situation for the working point of the SHG resonator. Moreover, assembly of the resonator was fast and flexible.

At the same time the resonator was very compact with a short arm length of

2. High-power 532 nm single-frequency TEM$_{00}$ laser sources

2.3 cm between the plane mirrors and a long arm length of 7.9 cm between the curved mirrors. The two parallel arms had a distance of 4.4 cm. These dimensions resulted in an optical round trip length of only 25.3 cm with the nonlinear crystal in place, corresponding to an $FSR = 1.18$ GHz. After optimization the input coupler IC had a measured transmission for the fundamental light of $T_{in} = 8.1\%$, the other mirrors were highly reflective with only spurious transmission at a wavelength of 1064 nm. Mirror DC had a dichroic coating, which was highly reflective for the fundamental light but transmitted 95 % of the harmonic light at a wavelength of 532 nm. The curved mirrors had a radius of curvature of 250 mm and formed a waist of 220 μm at the center of the 2 cm long PPKTP crystal with facet dimension of 1 mm × 2 mm. The waist showed no measurable astigmatism or ellipticity.

To allow for such a compact resonator design, the crystal oven had to be rather compact, too. It used small Peltier elements to keep the PPKTP crystal at its phase matching temperature of 38 °C inside its thermal confinement made from copper. An isolating shield made from POM protected the crystal from temperature disturbances from the ambient conditions. To optimize the thermal contact between crystal and copper, the copper facets were polished and the crystal was wrapped into thin indium foil. Indium foil was also put on both sides of the Peltier elements. Additionally, the crystal, the copper parts and the Peltier elements were clamped from above by its isolating hull with sufficient force to deform the indium. This was achieved with the help of small angled wings at the hull's outer side. These wings were pushed down by screws. For even more compactness the crystal was placed off-center and the front and back isolation had recesses for the angled beams of the bow-tie resonator. Thin aluminum covers protected the isolation material from wandering beams. A design drawing of the oven is shown in Fig. 2.8.

The stabilization of the oven temperature was done by a feedback control loop which read out the voltage drop over an NTC, which was glued into a small drilling in the copper parts of the oven. The control loop's unity gain frequency (UGF) was 0.4 Hz and temperature stabilities of 10 mK were routinely achieved when no large optical power changes occurred inside the crystal.

The stabilization of a resonance frequency of the SHG resonator to the frequency

2.3. Converting an intermediate power metrology laser

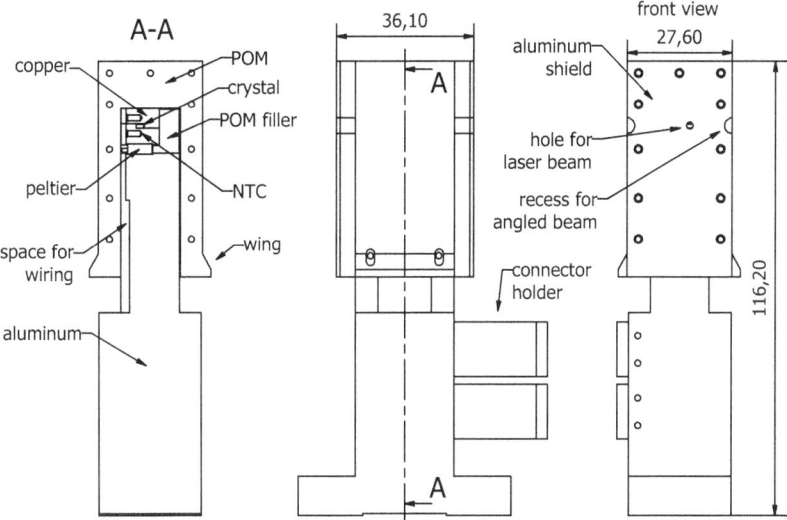

Fig. 2.8: Design drawing of the self-made compact oven, which was used for the temperature stabilization of the PPKTP crystal. All given length values are in units of millimeters. The crystal itself was wrapped into indium foil and placed between the two copper parts.

of the incident light was done via a modulation/demodulation technique similar to the Pound-Drever-Hall locking scheme [22]. In order to remain independent from the availability of a phase modulation device in front of the SHG resonator or as part of the primary laser, the modulation sidebands were imposed onto the circulating light inside the SHG cavity by the same piezo-electrically adjustable mirror, which was described above. This technique is sometimes called dither lock [79]. The modulation frequency was approximately 1.5 MHz and thus created sidebands, which lay well inside the linewidth of the optical resonator of $FWHM \approx 8$ MHz. Although such a frequency is orders of magnitude larger than the first mechanical resonance of the piezo-driven mirror, still a considerable number of easily observable mechanical resonances with widths on the order of 50 kHz could be found in the demodulated signal for modulation frequencies between 900 kHz and 3 MHz. The modulated signal was sensed after leaving the resonator with photodetector PD3. It was equipped with an additional readout path, which was made resonant at the modulation frequency actually chosen to improve the signal-to-noise ratio (SNR). Then its output was demodulated, which resulted in an error signal with an SNR bigger than about 20 for an root mean square (RMS) modulation voltage of about 100 mV at the piezo. The demodulated signal was shaped by a proportional-integral-derivative (PID) controller and amplified within a dynamic range of $0 - 300$ V prior to feeding it back to the piezo driven mirror. The modulation signal was added capacitively to this high-voltage signal. This control loop had a UGF of approximately 10 kHz limited by the first mechanical resonance of the piezo-driven mirror. It was made independent from the presence of an experimenter by adding automatic lock-acquisition electronics to the loop, which reinitiated a locked state within approximately 2 s after a drop of the harmonic power on photodetector PD4. When these electronics were active the system was stabilized during 95 % or more of its operational time.

All four photodetectors utilized semiconductors as light-sensitive elements and had bandwidths on the order of 1 MHz. Their output voltage was calibrated to represent the optical power at calibration points CP1 to CP4 respectively. The uncertainty of this calibration was determined to be ± 3.7 % dominated by the accuracy of the power meter applied for the calibration.

2.3.5. Results and discussion

Initial performance

In the first weeks after its assembly the resonant PPKTP SHG stage showed remarkably good performance. This is presented in the following.

2.3. Converting an intermediate power metrology laser

Harmonic power and conversion efficiency. To determine the conversion performance of the resonant PPKTP SHG stage its harmonic output power at calibration point CP4 and its external conversion efficiency from CP1 to CP4 were measured. The result is presented in Fig. 2.9. A harmonic output power at CP4 of $P_\text{h} = 6.9\,\text{W}$ was stably produced when a fundamental power at CP1 of $P_\text{f} = 10.2\,\text{W}$ was incident on the SHG resonator, corresponding to an external conversion efficiency of 68 %. This external conversion efficiency was defined as

$$\eta_\text{SHG} = \frac{P_\text{h}(\text{CP4})}{P_\text{f}(\text{CP1})} \qquad (2.37)$$

The highest value of $\eta_\text{SHG} = 69\,\%$ was achieved at a harmonic power of 5.6 W. The error bars reflect the calibration uncertainty given in the experimental setup subsection above. To the best of the author's knowledge this result was among the highest optical powers at a wavelength of 532 nm ever reported for SHG in a PPKTP crystal (see also discussion in 2.3.1). The measured power levels remained stable in time and reproducable within their errors for at least 30 h of full-power operation.

Importance of losses and damage threshold. Fig. 2.9 compares the measured data with the prediction of the model developed in 2.2.5, which assumes an ideal nonlinear crystal, that is not influenced by heating from optical absorption. The values of the model parameters are mostly those of Tbl. 2.1 already used for the design simulation of the system with the exception that the transmission of the input coupler was now measured to be $T_\text{in} = 8.1\,\%$ and the passive roundtrip losses were determined to be $A_\text{p} = 1.5\,\%$. For harmonic powers above ca. 3 W there is an increasing discrepancy between the model prediction and measurement such that the model predicts a higher harmonic power than is actually measured. For highest powers this discrepancy grows to 10.1 % of the measured harmonic power. Thus the assumption of such an ideal nonlinear crystal without heating by absorption seemed to be invalid for the PPKTP crystal of the SHG resonator.

In Chapter 3 an identical resonant SHG stage is described, which was equipped with a more powerful infrared laser. For comparable power levels the SHG performance in that case was essentially identical to this system. However, when testing

2. High-power 532 nm single-frequency TEM$_{00}$ laser sources

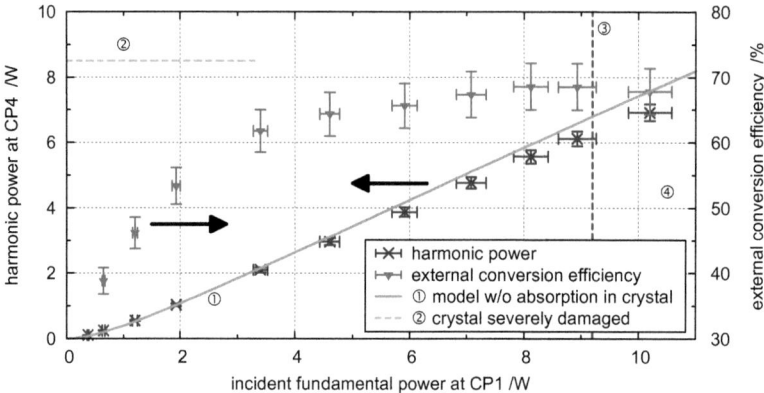

Fig. 2.9: Conversion performance of resonant PPKTP SHG stage in the first weeks after assembly compared with a model, which neglects absorption inside the crystal. Incident powers in the yellow region (4) caused the frequency control loop to become increasingly instable. The gray vertical line (3) depicts the incident power level, at which the lateral harmonic beam shape was measured (see Fig. 2.10). At the power level marked by the horizontal broken line a sudden crystal damage occurred in an identical SHG stage with a more powerful infrared laser.

even higher incident powers the lock acquisition and stabilization of the SHG resonator became more and more difficult for incident powers above 10 W (marked as yellow region in the figure). No obvious reason for this could be found in the frequency stabilization control loop. Instead, it seemed that fluctuations in circulating power, crystal temperature and resonance frequency became increasingly coupled, which caused the overall system to become instable. In a similar manner like the harmonic power discrepancy of Fig. 2.9 this increasing instability also hinted on absorption in the nonlinear crystal to be important. At a harmonic power of approximately 8.5 W a sudden damage occurred to the crystal, which caused an irreversible drop

2.3. Converting an intermediate power metrology laser

of the maximum harmonic output power to approximately one third of its previous value and a strong increase in crystal absorption. This power level is also indicated in Fig. 2.9 by the broken horizontal line in the left half.

Harmonic beam shape. If the harmonic power should be used for precision metrology experiments or if it should be enhanced inside another optical resonator, it is often important to have a high beam quality, meaning that the fraction of harmonic power η_{00} contained in the TEM$_{00}$ mode should be as high as possible. Examples can be found in literature, where absorption inside the crystal or other effects distort the lateral beam shape and thus strongly reduce this fraction [78, 142, 111]. The lateral distribution of the harmonic intensity was measured at high harmonic power (marked in Fig. 2.9 by the vertical gray line) to obtain an estimation of its beam quality. The measurement device was a WinCam charge-coupled device (CCD) camera. The result is shown in Fig. 2.10. Obviously the harmonic beam showed a high quality, proven by a circular intensity distribution with a Gaussian fit quality of $G = 96.6\,\%$ on average. This fit quality is defined as [37]

$$G = 100\times \left(1 - \frac{\text{sum of absolute differences}}{\text{Gaussian profile area}}\right) \quad .$$

Nevertheless a very weak ring structure was recognizable around the Gaussian part, which hinted on a shift of several percent of the beam's power into higher-order transversal modes. Movement and tilt of the filters and the CCD camera itself during this measurement caused the narrow stripes and small ring-like interference patterns in the data to move and partly vanish. From this it was concluded that they were measurement artifacts and not part of the beam's intensity.

Modelling of SHG resonator with absorbing crystal. As was found above the measured harmonic output power could not be well described by a model, which did not account for heating of the nonlinear crystal by optical absorption. In 2.2.5 an extension to the SHG resonator model was derived, which does account for such effects by assuming that heating by optical absorption (linear and nonlinear) causes thermal dephasing, i.e. a spoiling of the phase matching condition as a function of

2. High-power 532 nm single-frequency TEM$_{00}$ laser sources

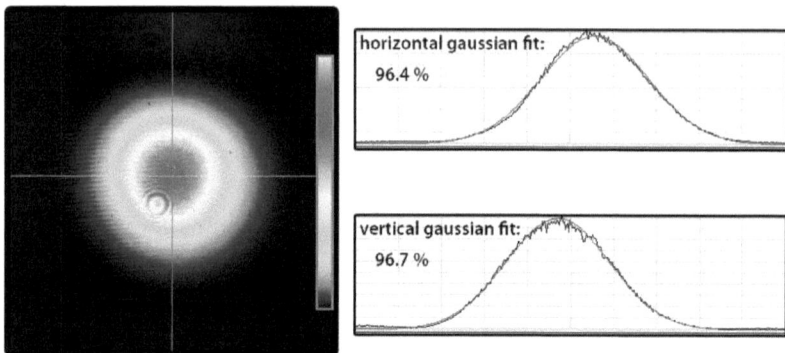

Fig. 2.10: Lateral distribution of the harmonic intensity generated by the resonant PPKTP SHG stage at a harmonic power of 6.3 W. The left part shows a two dimensional measurement with a CCD camera. The color code is given by the vertical linear color bar on the right edge of the left part. The higher the color in the bar, the higher was the detected intensity. The right part shows Gaussian fits to the data along the white intersection lines in the left part. The fit quality is given as the value of G defined in the text. The narrow stripes and ring-like interference patterns were measurement artifacts and not part of the beam's intensity. A weak outer ring structure is recognizable (sometimes badly visible in printed versions, please refer to online version).

the lateral position (see Eq. (2.36)). In this model extension the absorption coefficients are initially free parameters, which can either be measured directly by some accompanying experiment or are determined indirectly from a fit to the measured performance data of a resonant SHG stage.

To get an independent measurement of the absorption coefficient for fundamental light, an identical PPKTP crystal was traversed by a high-power TEM$_{00}$ laser beam at a wavelength of 1064 nm, which formed a radius of 76 µm close to the crystal's

2.3. Converting an intermediate power metrology laser

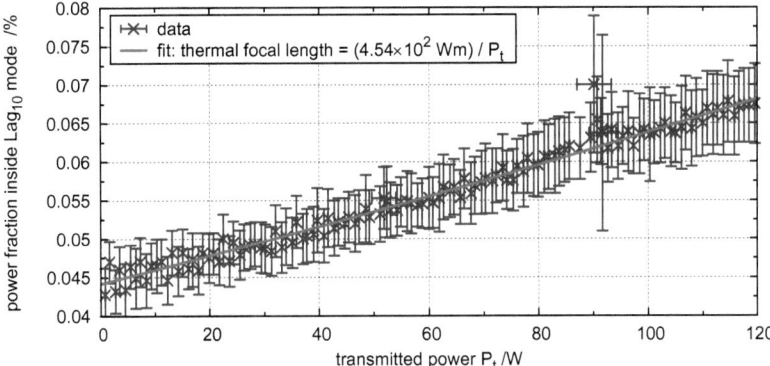

Fig. 2.11: Measurement of the power fraction of a TEM$_{00}$ laser beam at 1064 nm, which was transferred into the LG$_{10}$ mode by transmission through a PPKTP crystal far from its phase matching temperature. The crystal was mounted in the same way as for the resonant SHG stage. The beam was focussed to a waist size of 76 µm located 19 mm off the crystal center.

center. The crystal was identical to the one used in the resonant SHG stage (i.e. both were purchased together from the same manufacturer) and mounted in the same way, but kept far from its phase matching temperature such that no harmonic light was generated. After transmission of the near-infrared (NIR) beam the device described in [80] was used to determine the power fraction, which was shifted into the LG$_{10}$ mode by the thermal lens induced inside the crystal by heating through absorption. From this measurement the focal length f_{th} of this thermal lens was deduced. The result is shown in Fig. 2.11 [24]. However, neither the development of this measurement technique nor the experiment itself and its evaluation was done by the author of this thesis.

The resulting value for the thermal focal length for a transmitted power of 120 W was $f_{th} = 3.78$ m. The SNR appeared to be acceptable. Several models can be found

2. High-power 532 nm single-frequency TEM$_{00}$ laser sources

in the literature for the calculation of the expected focal length of a thermal lens for a given set of experimental and material parameters [71, 128, 94]. As the thermal lens in this case is very weak, a lowest order model should be sufficiently precise. Thus the equation given in [71] was adopted. The model equation reads

$$\frac{1}{f_{th}} = \frac{\alpha_{l,f} P_{inc}}{\pi \kappa_{th}} \left(\frac{\partial n}{\partial T} \right) \int_{z_0-L_k/2}^{z_0+L_k/2} \frac{1}{w^2(z)} dz \qquad (2.38)$$

$$= \frac{\alpha_{l,f} P_{inc} n_f}{\lambda_{0,f} \kappa_{th}} \left(\frac{\partial n}{\partial T} \right) \left[\arctan\left(\frac{z_0 + L_k/2}{z_r} \right) - \arctan\left(\frac{z_0 - L_k/2}{z_r} \right) \right] .$$

In this expression Eq. (2.5) was used for the beam radius $w(z)$ and z_0 denotes the distance of the waist from the crystal center. The prediction for f_{th} was plotted over the linear fundamental absorption coefficient $\alpha_{l,f}$ in Fig. 2.12 to find that value of $\alpha_{l,f}$, which causes a thermal lens of the measured size. The material parameters were $\kappa_{th} = 3.0 \frac{W}{Km}$, $n_f = 1.74$ and $\partial n_f / \partial T = 7 \times 10^{-6} \frac{1}{K}$, with the refractive index and its temperature dependence for light polarized perpendicular to the optical z-axis as was applied in the thermal lens measurement [92]. The resulting linear absorption coefficient of the PPKTP crystal for the fundamental light was $\alpha_{l,f} = 1.2 \times 10^{-3} \frac{1}{m}$. This is a remarkably low value for a periodically poled crystal. Even high quality Suprasil (i.e. fused silica) substrates usually show absorption coefficients on the order of $\alpha_{l,f} \approx 2 \times 10^{-4} \frac{1}{m}$ [89].

The absorption coefficient for the harmonic light could not be measured because a combination of laser source and measurement device for a wavelength of 532 nm was not available. Hence it was used as fitting parameter when modelling the harmonic power as a function of the incident power incorporating the measured value of $\alpha_{l,f}$. The effect of GRIIRA was assumed to be negligible as the harmonic intensity was reduced far below the relevant intensities already given in 2.2.4 by the realization of a large fundamental waist inside the crystal. The best fit of the thermal dephasing model to the data is shown in Fig. 2.13 and resulted in an absorption coefficient for the harmonic light of $\alpha_h = 2.4 \frac{1}{m}$. This value is much higher than the absorption coefficient for the fundamental light, which is in agreement with the range of values reported by other authors and also given in 2.2.4. The thermal dephasing model

2.3. Converting an intermediate power metrology laser

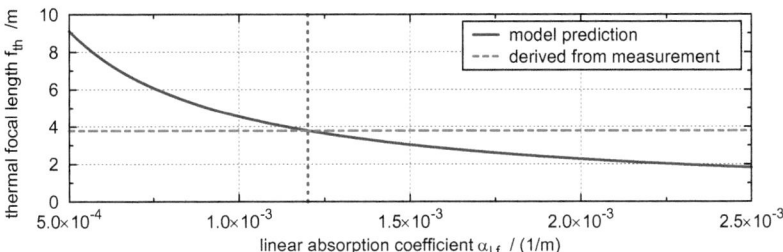

Fig. 2.12: Comparison of the measured value for the thermal lens in PPKTP from absorption of 1064 nm light with the prediction from a model. The model matches the measured value for an absorption coefficient of $\alpha_{l,f} = 1.2\times10^{-3}\,\frac{1}{\mathrm{m}}$, indicated by the vertical dotted gray line.

predicts the harmonic power significantly better than the model without absorption inside the nonlinear crystal.

If thermal dephasing is an issue then there is an interesting difference between a resonant SHG stage and a single-pass stage. In the single-pass case the harmonic power would be reduced by two effects, namely the loss of harmonic power due to absorption and the loss of single-pass conversion efficiency due to thermal dephasing. This is different in the case of a resonant stage for the same reasons, which allow for a larger fundamental waist in this case. As the single-pass conversion efficiency acts as a loss for the optical resonator, its power buildup increases when thermal dephasing takes place. The power buildup thus compensates largely for the loss in single-pass efficiency, until thermal dephasing due to fundamental absorption or passive absorption in the resonator becomes a limiting factor. This effect of rising circulating power when thermal dephasing takes place could be measured in the case of the resonant SHG stage described in this section and is shown in Fig. 2.14. The figure shows the measured circulating power at CP3 and the predictions of two models, one of which neglects absorption inside the crystal and the other includes thermal dephasing processes. Similar to the harmonic power also the circulating power is

2. High-power 532 nm single-frequency TEM$_{00}$ laser sources

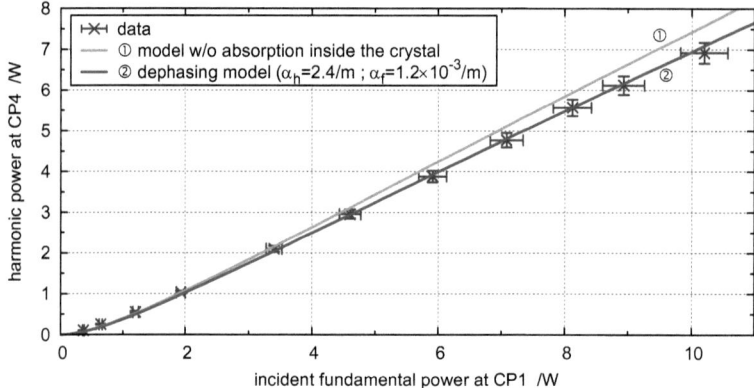

Fig. 2.13: Harmonic output power of the resonant PPKTP SHG stage in the first weeks after assembly compared with a model, which assumes no absorption inside the crystal, and with a model, which incorporates thermal dephasing processes inside the nonlinear crystal caused by absorption from fundamental and harmonic beam. Incident powers in the yellow region caused the frequency control loop to become increasingly instable.

better described by the dephasing model, which is clearly visible for those incident power levels, where thermal dephasing lowers the harmonic power in Fig. 2.13.

All in all the conversion performance of the resonant PPKTP SHG stage could be well modelled with the set of absorption coefficients given in Fig. 2.14 as parameters of the thermal dephasing model based on Eq. (2.36). The fitted value of the harmonic absorption coefficient lies well in the range of values reported in literature. The one of the fundamental absorption coefficient, which was derived from an independent measurement is remarkably low (see 2.2.4 for a compilation).

As a final consideration it was interesting to check if the chosen transmission of the input coupler in fact realized an impedance matched resonator. This was done by calculating the fraction of incident power, which was reflected from the input

2.3. Converting an intermediate power metrology laser

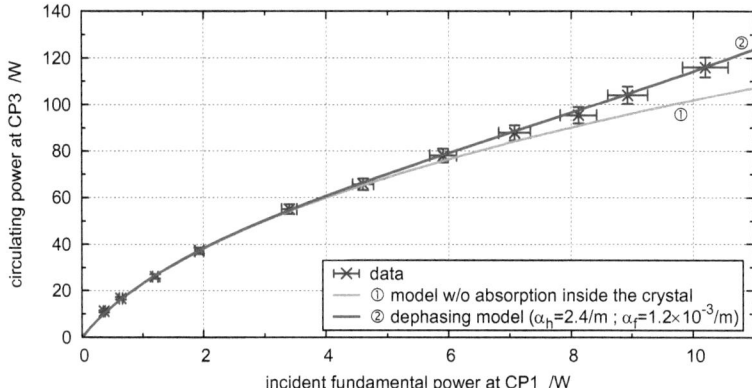

Fig. 2.14: Circulating fundamental power inside the resonator of the PPKTP SHG stage in the first weeks after assembly compared with a model, which assumes no absorption inside the crystal, and with a model, which incorporates thermal dephasing processes inside the nonlinear crystal caused by absorption from fundamental and harmonic light.

coupler $P_{\text{refl}}(\text{CP2})/P_{\text{inc}}(\text{CP1})$. The result is shown in Fig. 2.15 again together with the same models as in the figures above. Obviously the reflected power fraction did not follow the models, which in fact both predicted a nearly impedance matched SHG resonator for maximum harmonic power. Instead it settled to a reflected fraction of ca. $(1 - \eta_{00}) = 7.9\,\%$. This fraction was deduced to originate from a mismatch between the eigenmode of the SHG resonator and the incident beam. Careful alignment did not reduce it. As this power fraction never entered the SHG resonator on its TEM_{00} resonance, the effective incident power for the calculation of the models has already been corrected for this fraction in all figures above.

The remaining mismatch between the measured data and the model including imperfect matching of the modes in Fig. 2.15 is unknown. But it might be explained by the occurrence of thermal lensing inside the nonlinear crystal. If strong enough

2. High-power 532 nm single-frequency TEM$_{00}$ laser sources

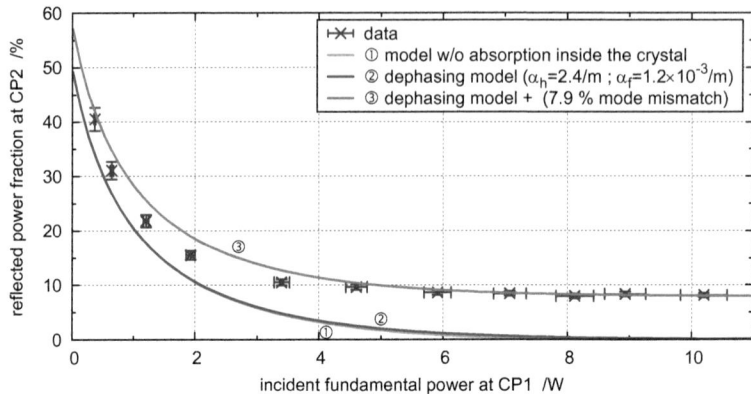

Fig. 2.15: Fraction of incident power reflected from the resonator of the PPKTP SHG stage in the first weeks after assembly. Also shown are the prediction of a model, which assumes no absorption inside the crystal, and of a model, which incorporates thermal dephasing processes inside the nonlinear crystal caused by absorption from fundamental and harmonic beam. The mismatch between incident and resonator eigenmode was derived from the remaining reflected power fraction in a nearly impedance matched situation and added to the dephasing model.

this would change the mode shape of the resonator with varying circulating (and hence also incident) power. This effect cannot be predicted correctly by the applied model.

Very slow performance degradation

Settled performance. The performance of the resonant PPKTP SHG stage described above remained stable and reproducible within the measurement uncertainties for approximately 30 h of operation close to its maximum harmonic power. On

2.3. Converting an intermediate power metrology laser

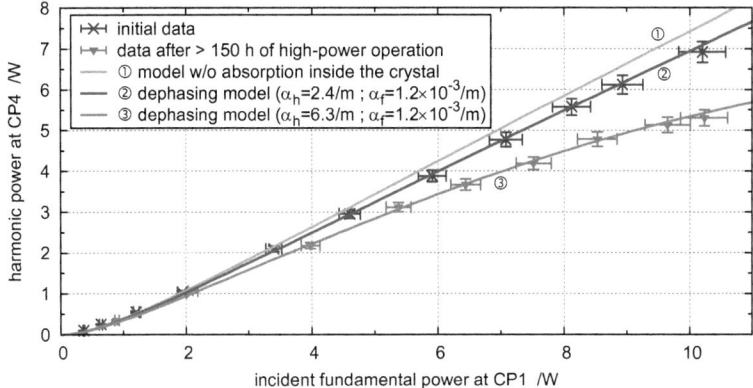

Fig. 2.16: Comparison of the harmonic output power of the resonant PPKTP SHG stage at its initial state and after approximately 150 h of operation close to maximum harmonic power. Also shown are one prediction from a model, which assumes no absorption inside the crystal, and two from a model, which incorporates thermal dephasing processes inside the nonlinear crystal caused by absorption from fundamental and harmonic light. Incident powers in the yellow region caused the frequency control loop to become increasingly instable.

longer timescales a very slow performance degradation became observable. After a certain period, whose length could only be roughly estimated to be on the order of 150 h, this slow performance degradation appeared to have settled to a considerably lower power level. The diminished harmonic power level is presented in comparison with the initial one in Fig. 2.16. This power level then remained stable over more than 300 h of operation. Accordingly, this experiment generated, to the best of the author's knowledge, the highest long-term stable harmonic power of single-frequency 532 nm laser radiation obtained from PPKTP, which was published so far. It amounted to 5.3 W.

2. High-power 532 nm single-frequency TEM$_{00}$ laser sources

As was described in 2.2.4 the slow degradation process might have been caused by the effect of gray tracking, which denotes a time-dependent saturating degradation in nonlinear crystals. Its time constant depends on the harmonic intensity, which was strongly reduced in this setup by the choice of a particularly big fundamental waist. Accordingly one would expect either a non-existent gray tracking process as was implied for instance by [84], or the process should be very slow, which would be the implication of the results shown here. If gray tracking was the reason here, one would expect the harmonic absorption coefficient α_h to have risen, while the fundamental one should have remained the same. Fig. 2.16 also shows a fit of the thermal dephasing model to the data, where only α_h was allowed to change in comparison with the parameter set, which successfully modelled the initial phase of the experiment. The fit resulted in a harmonic absorption coefficient of $\alpha_h = 6.3\,\frac{1}{m}$ and reproduced the measured data well. Thus the value of α_h seemed to have risen by 4.2 $\frac{1}{m}$ over long timescales. In single-pass experiments by other authors a much higher increase by 13 $\frac{1}{m}$ was found on much shorter timescales of only a few hours [84]. The considerable improvement presented here might be related to a much higher crystal quality or to the by far lower intensities used in this setup.

Dissipated power. As an independent test of the dephasing model applied above, the electrical power dissipated in the crystal oven was measured when the SHG stage was operated at a harmonic power level of 5.0 W in its settled state. As the phase matching temperature of 38 °C was well above ambient temperature the Peltier elements of the crystal oven were in heating mode when the laser was off as well as in its 'on' state at the given harmonic power level. In contrast to the case of cooling Peltier elements, the heat generated by a chain of them in heating mode should be given simply by the electric power dissipated in them. The controller of the temperature stabilization loop of the PPKTP oven allowed to measure this dissipated electrical power. When the circulating fundamental power was switched from zero to approximately 110 W, which corresponded to the harmonic power level just given, a change in dissipated electric power of 480 mW was measured after two minutes. Obviously the missing electric heating power in the 'on' state was provided

2.3. Converting an intermediate power metrology laser

by the fundamental and harmonic beams in the crystal via optical absorption. The value of α_h for the settled case presented in Fig. 2.16 implies that for a harmonic power of 5 W a heating power of 590 mW was generated inside the crystal while the heating effect from the circulating fundamental light was negligible. Thus there is at least a qualitative matching of the reduction in electric heating power and of the increase of heating from optical absorption. This strengthens the validity of the thermal dephasing model applied above. The discrepancy should arise from the simplifying assumptions made in the development of the dephasing model, e.g. the negligence of stress-induced birefringence from the thermal gradient.

Temperature gradient. The fits of the dephasing model to the initial and settled SHG performance in Fig. 2.9 and Fig. 2.16 allowed to calculate the evolution of the radial temperature gradient inside the crystal between the beam's center and the waist of the fundamental beam for the two cases. Both are presented in Fig. 2.17 and were normalized to the thermal acceptance bandwidth δT_{bw} of the PPKTP crystal to demonstrate its importance. Comparison of this result with Eq. (2.25) and Eq. (2.26) shows that one would expect a remarkable difference between the optimum phase matching temperatures for the SHG cavity kept on resonance and for the case in which its length is ramped.

Such a difference was indeed measured for this setup in its settled state. If the resonator is ramped and the fundamental airy peaks are monitored for both directions of the ramp, each of them is expected to be symmetric for the case of optimum phase matching. In this case the conversion process is simply a loss for the circulating fundamental light but nothing else happens. If the phase matching temperature is slightly detuned, the cascaded Kerr effect will become important in addition to the conversion loss (a detailed description of this effect is beyond the scope of this thesis; see for instance [138, 147] for a detailed explanation). The only aspect of this effect, which is important here, is the following. With increasing detuning of the phase matching temperature the airy peaks will become increasingly asymmetric, leaning a bit to the side. In contrast to effects caused by bandwidth limitations of any kind, the cascaded Kerr effect will cause airy peaks of opposing ramp directions

2. High-power 532 nm single-frequency TEM$_{00}$ laser sources

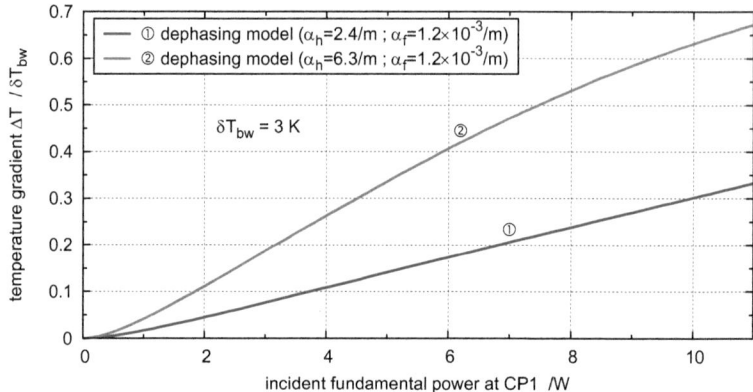

Fig. 2.17: The temperature gradient from the beam's center to the fundamental waist inside the PPKTP crystal as derived from the dephasing model for the values of the harmonic und fundamental absorption coefficients derived in this section. The temperature was normalized to the thermal acceptance bandwidth of the PPKTP crystal used in the resonant SHG stage.

to be inclined into *opposing* directions [138]. As long as the inclinations are small enough not to cause a bistability of the resonator, the harmonic airy peaks are expected to behave essentially identical. Turning this argument around clarifies that if one measures such an inclination of the harmonic airy peaks, one can deduce a non-optimal phase matching temperature of the conversion process.

The upper graph of Fig. 2.18 presents measurements of the harmonic airy peaks for two opposing ramp directions. The phase matching temperature of the PPKTP crystal was the only parameter that was changed between the three displayed airy peaks per ramp, especially the speed of the ramp was the same for all measurements. The symmetric harmonic airy peak was obtained, when the phase matching temperature was optimized for the given ramp speed. Slight detuning of the phase

2.3. Converting an intermediate power metrology laser

matching temperature resulted in the inclined but undistorted peak. If, however, the phase matching temperature was adjusted such that the harmonic power was maximized to about 5 W in a *stabilized* (i.e. static) situation, then in the ramped case massively distorted harmonic airy peaks were obtained. The massive distortion arises from discontinuous jumps of the harmonic power caused by ramping a bistable optical resonator. Each discontinuous jump is then distorted by bandwidth limitations of the photodetector. Such a bistability of the nonlinear resonator implies an especially strong detuning of the phase matching temperature.

Obviously the optimum phase matching temperature differed a lot between a static situation with high and constant harmonic power and a ramped situation with low harmonic power on average. Such a behavior would be expected from the plots in Fig. 2.17 and therefore this result again strengthens the validity of the dephasing model applied above.

Summary of the PPKTP SHG experiment. Although a long-term stable harmonic power of 5.3 W from PPKTP is a success, it has become clear in this section, that this material cannot be the choice for significantly higher power SHG stages, unless crystals of by far superior quality become available. The current quality levels result on the one hand in too high linear absorption coefficients, especially for the harmonic light. On the other hand on very long timescales also significant gray tracking takes place, although the harmonic intensity was reduced to very low levels.

Since the time when the original version of this experiment was assembled, most experimenters in the world decided rather to use PPSLT if a high nonlinearity material was necessary. This material has a higher thermal conductivity and purity and hence allowed higher harmonic output powers, while showing a nonlinearity comparable in size to PPKTP (e.g. see [132, 124]). If a low nonlinearity material is sufficient and if high harmonic powers are intended, focussing on materials with very low linear and nonlinear absorption is expected to be most promising.

2. High-power 532 nm single-frequency TEM$_{00}$ laser sources

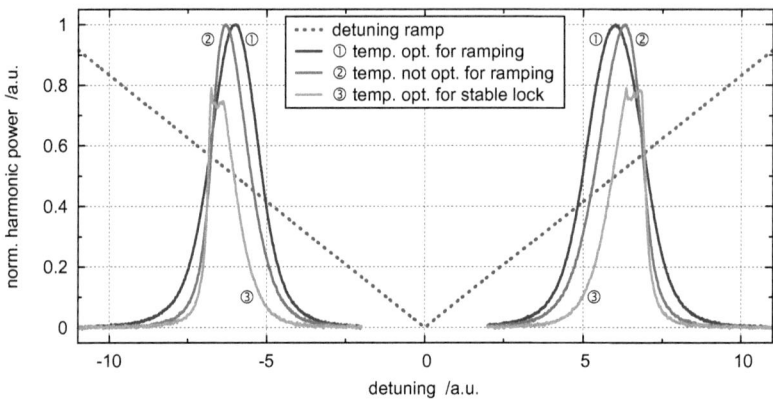

Fig. 2.18: Measured harmonic airy peaks when the phase matching temperature was optimized for ramping, slightly detuned from that, and when it was optimized for a stabilized (i.e. static) SHG resonator with harmonic power of 5 W. Here, optimization means maximization of harmonic peak power. In all cases the curves were scaled in vertical direction for better visibility. The distortion in the latter case is due to a bistability of the SHG resonator.

2.4. A 130 W CW single-frequency TEM$_{00}$ green laser source

With the development of the prototype of the Pre-Stabilised Laser (PSL) for the Advanced Laser Interferometer Gravitational-Wave Observatory (LIGO) an NIR CW laser system became available, which extraordinary well combined an extremely high single-frequency output power of 150 W with all properties necessary for a precision metrology laser. The laser light at a wavelength of 1064 nm was available downstream of the spatial filter cavity of the PSL. Its higher-order transversal mode

2.4. A 130 W CW single-frequency TEM$_{00}$ green laser source

content was measured to be less than 2 % [151]. Its frequency noise was comparable to that of a non-planar ring oscillator as in [81]. Its relative power noise usually remained below a few percent within minutes and pointing was below significance for the experiments described here.

For a gravitational-wave detector like Advanced LIGO such high laser powers are necessary to achieve the desired sensitivity but they are also a source of persistent problems, which originate from thermal effects from optical absorption and from stray light. Therefore lowering the optical power would be desirable for future generations of gravitational wave detectors. In general it is possible to do this without reduction of the detector's quantum noise limited sensitivity by a transition to shorter wavelengths. In fact the signal-to-quantum-noise ratio remains constant if both the optical wavelength and the circulating optical power are halved [23]. If the detector is limited by technical noise sources, which scale less than linear with the optical power, its sensitivity would be even enhanced. Plans for future space-borne missions like DECIGO or BBO already incorporate these ideas (see [52] and references therein).

High-power precision laser sources emitting laser light at a wavelength of 532 nm are also important in other fields of science and engineering as was explained in the introduction to this chapter. Thus it was interesting to set up an external-cavity SHG stage to convert the Advanced LIGO PSL prototype to the green visible spectral region.

2.4.1. Highest reported harmonic power levels for green light

The various schemes to implement an SHG stage and suitable nonlinear crystal materials were already explained in 2.2.3.

The single-pass SHG experiments reported in literature so far mainly applied nonlinear crystals with high effective nonlinearity to achieve acceptable conversion efficiencies. Unfortunately, one can find many examples reporting that the conversion efficiency of all of these materials is strongly degraded by thermal de-phasing processes when exposed to incident fundamental powers of multiple tens of

2. High-power 532 nm single-frequency TEM$_{00}$ laser sources

watts [145, 134, 84, 90, 132, 140, 78, 45, 97, 120]. The individual thermal dephasing processes were described in 2.2.4.

For single-pass SHG Sinha et al. reported the generation of 18 W of single longitudinal and (nearly) single transversal mode (referred to as *single-mode* for simplicity) green light available downstream of the SHG with 24 % efficiency [132] and Samanta et al. reported the generation of 13 W of single-mode green light with 56 % efficiency [124], which are the highest values known to the author for a single-/multi-pass design and which were limited by thermal dephasing effects.

To the best of the author's knowledge, literature on intracavity doubling schemes reports that in the multi-longitudinal mode case a maximum emission of approximately 60 W was achieved [96], while single-mode emission was limited to less than 10 W by beam distortion or stability of the CW single-mode emission state. These limitations had various reasons [158, 95], one prominent example being the so-called 'green problem', which denotes a problematic coupling between SHG process and laser process [12]. As an additional note the CUBRILAS collaboration achieved a long-term stable multi longitudinal mode harmonic power of 200 − 250 W at a wavelength of 515 nm from a commercial welding laser. However, this result was not scientifically published and no further details are known to the author [34, 157].

Finally, for external-cavity doubling schemes reports on setups with high as well as low effective nonlinearity crystals can be found in literature. But again crystals with high nonlinearities suffered from problems, which appeared to be related to thermal dephasing [56, 115, 71, 139]. This was also demonstrated in Section 2.3 of this chapter for the case of an external-cavity SHG stage based on PPKTP.

In contrast, setups, which utilized LBO crystals (which show low nonlinearity) did not suffer from such problems and generation of more than 20 W of single-mode light at 532 nm and 50 W at 589 nm were reported in literature [136, 137]. Again these values are the highest reported values known to the author.

2.4.2. Choice of design and crystal

In general there are three low nonlinearity crystal materials, which are nowadays widely used and seem to be rather insensitive to high optical powers. These are LBO, with a typical effective nonlinearity of $d_\text{eff} \approx 0.8\,\frac{\text{pm}}{\text{V}}$ [104], BBO with $d_\text{eff} \approx 2\,\frac{\text{pm}}{\text{V}}$ [103] and BiBO with $d_\text{eff} \approx 3\,\frac{\text{pm}}{\text{V}}$ [43]. Based on the arguments given in 2.2.3 and the previous subsection the scheme of external-cavity doubling and the nonlinear material LBO were chosen for the frequency conversion of the Advanced LIGO PSL.

Additional reasons for the crystal choice were the following. The experiments aimed at the generation of a single-mode harmonic beam in the sense described in the previous subsection. Thus it is beneficial that LBO can be noncritically phase matched by temperature tuning, because that assures a collinear propagation of fundamental and harmonic beam inside the nonlinear crystal. In contrast, BBO and BiBO have to be critically phase matched, which introduces a walk-off angle between the two beams inside the crystal of more than 25 mrad, which severely limits the interaction lengths of the two beams and tends to distort the harmonic beam shape [110, 142]. Additionally, there is some literature, which hints on the possible onset of degradation of BiBO and BBO (e.g. [69, 36] and [136]) when exposed to high power levels of light in the visible spectral region. However, many publications can be found reporting that LBO shows no significant degradation even when exposed to high optical power levels of several tens of watts (e.g. [136, 153, 137]). Finally, BBO is very hygroscopic [43] but LBO is expected to be not [86]. If LBO would be slightly hygroscopic the impact on the material quality should be further reduced if the crystal is steadily kept at its high phase matching temperature of about 149 °C [155].

2.4.3. Design of the SHG experiment

The general design considerations listed in 2.3.2 were applied in the planning phase of this experiment, too. First the effective nonlinearity of the LBO crystal was determined by fitting Eq. (2.27) to the result of a single-pass SHG experiment. This evaluation is shown in Fig. 2.19. An effective nonlinearity of $d_\text{eff} = 0.704\,\frac{\text{pm}}{\text{V}}$ was

2. High-power 532 nm single-frequency TEM$_{00}$ laser sources

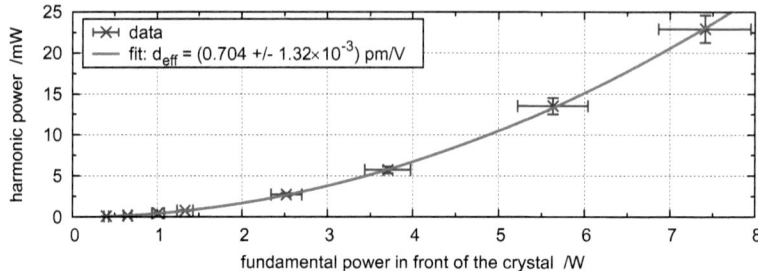

Fig. 2.19: Measurement of harmonic power obtained from single-pass of laser beam with waist of $w_{0,f} = 78\,\mu\text{m}$ at center of LBO crystal used for resonant SHG. The crystal's effective nonlinearity d_{eff} was determined from a fit of Eq. (2.27) to the data.

obtained.

In analogy to the design procedure of the PPKTP SHG stage in the previous section the SHG resonator was simulated prior to assembly. This was done to find a suitable fundamental waist size inside the crystal leading to reduced fundamental and harmonic intensities. The model was the same as was applied in the previous section (see 2.2.5). It neglected thermal dephasing caused by optical absorption inside the LBO crystal. The estimated and measured design values for the various parameters are given in Tbl. 2.2. The availability of very long crystals was another advantage of LBO. The value for the passive fractional round trip losses was taken from the one determined for the PPKTP SHG stage because optics of similar quality were used for this experiment.

The results of the simulation are shown in Fig. 2.20 and their qualitative behavior was similar to that obtained for the PPKTP SHG. However, the range of fundamental waist sizes, over which the reduction of the expected harmonic power is negligible is smaller in this case due to the smaller value of d_{eff} relative to the passive losses. The waist size, which corresponds to Boyd-Kleinman focussing is now $w_{0,f} = 43\,\mu\text{m}$ because the LBO crystal is much longer than the PPKTP crystal.

2.4. A 130 W CW single-frequency TEM$_{00}$ green laser source

parameter	value
L_k	5.0 cm
n	1.60
$\lambda_{0,f}$	1064 nm
P_{inc}	150.0 W
T_{dc}	97.5 % at 532 nm
A_p	1.5 %
d_{eff}	0.704 $\frac{pm}{V}$
η_{00}	95 %
T_{in}	imp. matched

Table 2.2: List of estimated as well as measured design values for experimental parameters relevant for the simulation of the resonant LBO SHG stage. The values of P_{inc}, A_p and T_{in} are only estimations here. Their measured values are given further below.

As the phase matching temperature of the LBO crystal is very high the dimensions of its oven were quite big to allow for thermal isolation (a rectangular region of about 52 mm in width and 72 mm in length was necessary). These dimensions considerably limited the flexibility of the design of the resonator's eigenmode and correspondingly the achievable fundamental waist sizes. Hence in this case only a comparatively small increase of the fundamental waist size by a factor of ca. 1.5 compared to the Boyd-Kleinman focussing regime was obtained from the iterative design procedure described in 2.3.2. Accordingly, a fundamental waist of 65 µm was chosen, corresponding to intensities at the crystal center of 9.2 $\frac{MW}{cm^2}$ for the fundamental and 3.9 $\frac{MW}{cm^2}$ for the harmonic. The highest intensities on any coated facet inside the resonator were then located at the crystal facets and amounted due to the diffraction of the beams to 3.6 $\frac{MW}{cm^2}$ for the fundamental beam and 1.5 $\frac{MW}{cm^2}$ for the harmonic beam. These intensities were close to the laser-induced damage thresholds stated for the employed optics (see 2.3.2). As further increase of the

2. High-power 532 nm single-frequency TEM$_{00}$ laser sources

fundamental waist size could not significantly reduce the intensities at the facets due to the beams' diffraction these values were taken as a given.

The expected harmonic output power was nearly not influenced at all by the waist enlargement. The expected round trip conversion loss for the fundamental light was determined from the simulation result to be 22 %, which had to be added to the expected passive roundtrip losses of 1.5 % to find the design value for the transmission coefficient of the input coupler IC of ca. 76 %.

2.4. A 130 W CW single-frequency TEM$_{00}$ green laser source

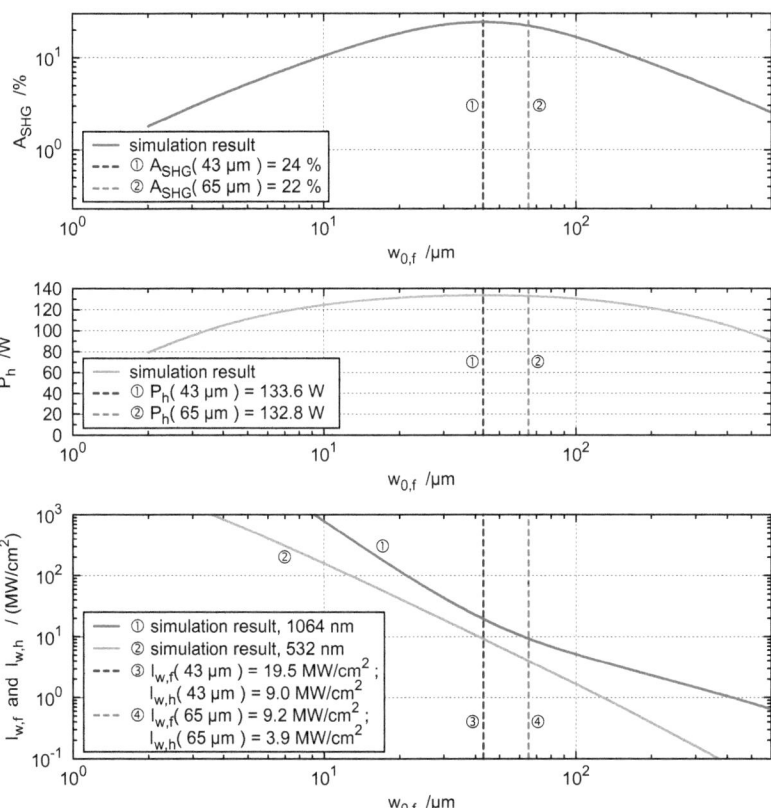

Fig. 2.20: Simulation of the LBO external-cavity SHG stage neglecting thermal dephasing (see 2.2.5 for the model and Tbl. 2.2 for the values of the parameters). Expected conversion losses (top), harmonic output power (middle) and peak intensities (bottom) are shown above the fundamental waist size. The vertical lines depict the waist size corresponding to Boyd-Kleinman focussing (blue, (1)) and the one chosen for this experiment (yellow, (2)).

2.4.4. Experimental setup

A schematic of the overall experiment is shown in Fig. 2.21. The infrared beam from the Advanced LIGO PSL was matched to the eigenmode of the SHG resonator by pairs of lenses and mirrors and its power could be adjusted with the help of a variable attenuator consisting of a $\lambda/2$-waveplate and a polarizing beam splitter. Its polarization was rotated by a second $\lambda/2$-waveplate to be parallel to the optical z-axis of the LBO crystal (which was vertical optical polarization) to obtain maximum conversion efficiency. A pick-off beam behind one turning mirror was used to measure the fundamental power incident on the SHG resonator with infrared detector CD1.

SHG resonator

The SHG resonator consisted of two convex (ROC=+1.0 m) and two concave mirrors (ROC=−0.2 m) in a bow-tie configuration. Convex mirrors were applied to enhance resonator design flexibility and thus to ease the incorporation of the rather big LBO oven. Mounting of the resonator mirrors was done in the same way as for the PPKTP SHG stage described in the previous section. The geometric lengths of the parallel resonator arms were 21.5 cm and 27 cm and their distance was 7.5 cm. The LBO crystal with dimensions 3 mm×3 mm×5 cm was mounted inside a customized version of commercial oven type KK1 from *EKSMA Optics* and placed in the longer resonator arm concentrically between the two concave mirrors. The oven was equipped with additional metal shields with a remaining clear aperture of 2.7 mm to prevent the secondary reflections of the circulating light from hitting teflon parts and it was controlled by *EKSMA Optics* commercial controller TK1. The crystal was cut for type 1 noncritical phase matching around approximately 149 °C. This resonator design resulted in an astigmatically compensated waist at the middle of the LBO crystal of 65 µm, which showed no measurable ellipticity or astigmatism. The second waist of the bow-tie resonator is denoted here as *coupling waist* and its average radius was 229 µm with a slight ellipticity of about 10 %. The amount of fundamental power circulating inside the resonator was detected by a

2.4. A 130 W CW single-frequency TEM$_{00}$ green laser source

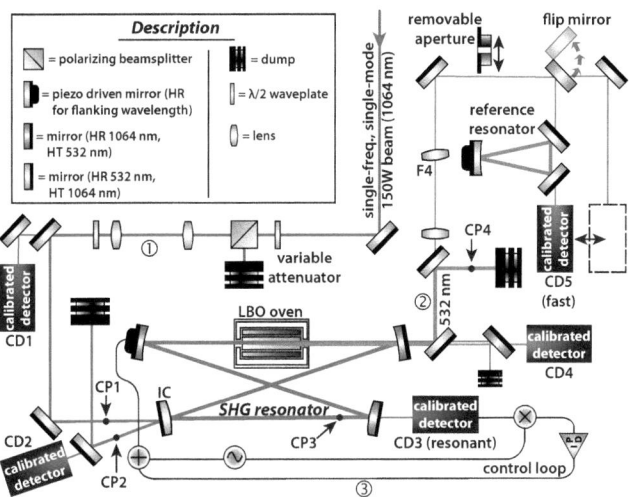

Fig. 2.21: Schematic overview of optical setup of LBO based resonant SHG stage. Red lines (1) depict fundamental infrared light, green lines (2) depict harmonic visible light, and black lines (3) denote electric wiring. *HR/HT xxx nm* denotes a mirror coating that is highly reflective or highly transmissive respectively for light with a wavelength of xxx nm.

pick-off beam, which originated from residual transmission of one resonator mirror and which was directed onto infrared detector CD3. The shape of the resonating fundamental mode could be observed with a CCD camera (omitted in Fig. 2.21), which detected a small fraction of the same pick-off beam. Most of the fundamental power reflected from the input coupler of the resonator (IC) was dumped, only a small amount was detected by infrared detector CD2. Similarly, most of the converted laser light at 532 nm was dumped, too, with the exception of two pick-off beams one of which was detected by visible detector CD4, while the other was used for modal analysis of the converted harmonic laser beam at 532 nm.

2. High-power 532 nm single-frequency TEM$_{00}$ laser sources

Mode analyzer

The polarization of that beam, which was intended to be used for the modal analysis, was adjusted by a $\lambda/2$-waveplate to be parallel to the table surface. Its mode was matched by a pair of lenses and mirrors to the eigenmode of a reference resonator, which consisted of a rigid aluminum spacer with three mirrors glued to it, one of them spaced by a piezo to allow for fast changes of the resonance frequency by more than two times its *FSR* of 714 MHz. The resonator finesse was 530 . The light transmitted by the ramped reference resonator was detected by the fast detector CD5. To gather additional modal information a black circular aperture of diameter $d_{ap} = 1.7$ mm could be placed in the beam at a distance of $z_{ap} = 40.5$ cm in front of the reference resonator waist of 263 µm. With the help of a flip mirror the light incident on the resonator could also be measured by CD5 directly. This whole set of devices for mode analysis of the harmonic beam will afterwards be referred to as *mode analyzer* and further explained in 2.4.5 and 2.4.6.

Detectors

The detectors CD1 to CD5 utilized photodiodes (made from InGaAs or silicon respectively) as light sensitive elements. Their response was assured to be linear over the dynamic range relevant in this experiment. By application of several filters it was assured that the detector's output signal was not influenced by light of undesired wavelengths or in the wrong polarization state. A lens as part of each detector assured the focussing of all incident light to the active area of the photodiode. The output voltage of the detectors CD1 to CD4 was calibrated by comparison with the respective optical power at points CP1 to CP4.

All calibrations and measurements were conducted only after the system had been in operation for at least an hour to allow it to reach a steady thermal state. The detector calibration was done by comparison of the detector output voltage on an oscilloscope with the power value measured by two different power meters (one for powers up to 10 W and one for powers up to 300 W), whose uncertainties were stated as 3 % (low power device) and 7 % (high power device). The oscilloscope DC

2.4. A 130 W CW single-frequency TEM$_{00}$ green laser source

gain accuracy was measured to be $\pm 1.5\%$. High frequency influences like electronic or laser power noise were determined from the measurements to be smaller than $\pm 1.6\%$. These uncertainties were expected to be independent from each other and correspondingly they were added quadratically. From this the maximum accuracy error of the detector calibration over the full dynamic range of the measurements was determined to be $\pm 7.3\%$. Detector output voltage and power measurement were compared for at least five points distributed over the relevant dynamic range. In all cases their relation was linear and the obtained calibration factors from both power measurement devices agreed well within their accuracy.

Coatings

Most resonator mirrors had a dichroic coating such that they transmit 97.5% of light at 532 nm but reflect more than 99.95 % at 1064 nm. This was done to maximize the external conversion efficiency and avoid unneccessary heating of the nonlinear crystal by circulating harmonic light. The only exception was the input coupler IC, which was only specified for 1064 nm and had a measured power transmission coefficient of $T_{in} = 25.5\%$. The LBO crystal was coated to reduce its reflectivity for both wavelengths to about 0.2 % per facet. To get rid off remaining infrared light in the beam path behind the SHG resonator dichroic mirrors were used, which reflected more than 99.8 % of the green light but less than 10 % of the infrared light. Unwanted infrared light in front of detector CD4 was attenuated further by introduction of another dichroic mirror with the same specifications as those forming the resonator.

Control scheme

The low reflectivity of the input coupler IC gave rise to a very broad linewidth of the SHG resonator, which eased the task of stabilization of its resonance frequency to the frequency of the incident laser light. The frequency stabilizations scheme itself was a copy of the one described in the previous section 2.3.

2.4.5. Conventional analysis of transverse mode structure

The analysis of the transversal mode structure of the harmonic beam was done by the *mode analyzer* as it is described in section 2.4.4. This device was used in three different operational modes to obtain an upper and two different lower bounds for the amount of harmonic power emitted into TEM_{00} mode. They are explained in the following.

Spectrum mode

The first operational mode will be denoted as *spectrum mode* (it is also known as mode scan technique) and it was adopted from [80]. It was realized in a rather similar way as is described in the following. First, the fraction of the beam under test, which was contained in the TEM_{00} mode, was matched to the optical resonator of the mode analyzer as well as possible. The resonator was ramped over a bit more than one *FSR* with a frequency of 250 Hz and the fast detector CD5 with a bandwidth of 5 MHz measured the transmitted power. Simultaneously detector CD4 measured the incident power. The resonator itself was constructed in a way, which assured that no degeneracy of resonances of transversal modes occurred for mode index sums below 30. Thus the transmitted power is a decomposition of the incident beam into its power fractions contained inside the various transversal eigenmodes of this resonator.

An electrical amplifier with two amplification stages with precisely trimmed gains of 10 and 100 and with flat bandwidths of 5 MHz was constructed. For the duration of a scan over the resonator's full *FSR* the two amplified output voltages, the not amplified one, and the incident power were recorded simultaneously via the four input channels of a special digital oscilloscope with a high horizontal resolution of 20000 points. Three such traces of four channels each were recorded for averaging purposes. Afterwards a computer algorithm first normalized the transmitted power to the incident one, then joined the three channels, which recorded the differently amplified transmitted signals, into a single one, and finally averaged over the three recorded traces after aligning their large TEM_{00} peaks. The result was a single

2.4. A 130 W CW single-frequency TEM$_{00}$ green laser source

mode scan with a high dynamic range of more than 2×10^3. Then another computer algorithm processed this high-resolution mode scan and identified as many higher-order transversal modes as possible. Finally, a model of an incident beam consisting of all identified higher-order modes was fitted to the mode scan to determine their precise peak heights. The integral incident power was determined from the sum of all these identified modes, and finally the power fraction η_{00} of the incident beam, which was contained inside the TEM$_{00}$ mode was calculated from this sum and the fitted height of the TEM$_{00}$ mode. Determination of the integral power in this way eliminated the need to know anything about the input coupler transmission or round trip losses of the optical resonator. The combination of the resonator finesse of 530 and the horizontal resolution of 20000 points assured that more than 30 data points lay within the *FWHM* of each mode's resonance, which was enough to result in a good fit.

The result of such a procedure at maximum harmonic power is presented in Fig. 2.22. The vertical resolution was limited by electronic noise. However, in any case the measurement will at a certain level be limited by the resolution of the ramped optical resonator, which is linked to its finesse. Thus the technique sketched above is in principle not able to exclude, that, in addition to all identified modes, the incident beam consists also of an infinite number of not identified higher-order modes with fractional powers just below the vertical resolution. This large amount of higher-order modes might for instance appear as a broad but arbitrarily weak corona around the beam like those known from amplified spontaneous emission processes. Therefore the amount of power contained in the TEM$_{00}$ mode obtained in the spectrum mode of the mode analyzer is strictly speaking just an upper limit on η_{00}. From the noise level in Fig. 2.22 the maximum power fraction in a single possibly not resolved higher-order mode can be conservatively estimated to be

$$\gamma_{\text{max}} \approx 5\times 10^{-4} \quad .$$

2. High-power 532 nm single-frequency TEM$_{00}$ laser sources

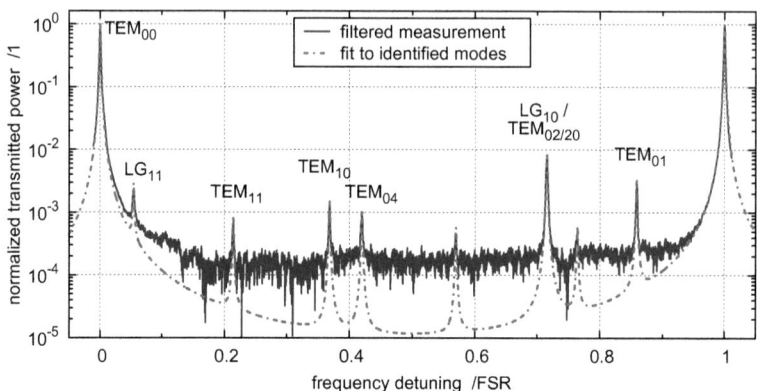

Fig. 2.22: High-resolution mode scan obtained by the spectrum mode of the mode analyzer from the power transmitted by the ramped mode analyzer resonator. Also shown is the fit to the identified higher-order transversal modes and the names of several modes, which were identified visually from CCD images. This result was obtained when the resonant SHG stage was set to its maximum harmonic power.

Transmission mode

To obtain also lower limits on η_{00}, the mode analyzer could be used in two additional operational modes. The first one was called *transmission mode*. This one is easily understood as it determined the amount of the incident power contained inside the TEM$_{00}$ mode simply by measuring the transmitted power fraction on the peak of the resonance of the TEM$_{00}$ mode. This was done by using the flip mirror in Fig. 2.21. It first diverted the light incident on the resonator and after that the light transmitted by it onto detector CD5, while the alignment of the incident beam to the resonator was kept the same as during the spectrum mode.

On the one hand this measurement will result in a true lower limit. But on the

2.4. A 130 W CW single-frequency TEM$_{00}$ green laser source

other hand it will in most cases clearly underestimate the value of η_{00}, because most optical resonators show a significant amount of impedance mismatch and round trip losses. These effects hinder a part of the incident TEM$_{00}$ power from traversing the resonator. Hence they reduce the power detected by CD5 and lower the derived value for η_{00}. They can easily account for several percent of the incident power.

A possible technique to determine a more stringent lower limit on η_{00} is presented in the following subsection.

2.4.6. Reduction of transverse mode analysis uncertainty

The mode analyzer could be used in yet another operational mode, called *aperture mode*. From this mode again a lower limit on η_{00} was obtained to narrow the discrepancy between the limits measured by the two operational modes described above.

Detector CD5 was used to measure the amount of light power incident on the resonator with and without a circular black aperture of diameter d_{ap}, which was placed symmetrically in the beam at a distance z_{ap} in front of the waist position inside the optical resonator. The alignment of the incident beam to the resonator was kept the same as in spectrum mode. The detector was placed right behind the aperture such that its collecting lens focussed all transmitted light, including any divergent parts from diffraction at the aperture edge, onto the detector's active area. The diameter of the circular aperture was narrowed down until a significant but very small power fraction L_{cut} was cut off by the aperture. To achieve a high measurement precision, a second detector was used, which measured the power incident on the aperture. Both detectors were not moved and the beam's alignment was not changed between the two measurements with and without aperture. Traces of length 10 s were recorded of the power in front and after the aperture to allow for normalization and averaging afterwards.

In 2.4.5 it was explained, that the result of the spectrum mode of the mode analyzer gives only an upper limit for η_{00}, because the incident beam might consist of a large number of additional not identified higher-order transversal modes. The

2. High-power 532 nm single-frequency TEM$_{00}$ laser sources

aperture mode is used to derive an upper limit on the power in these not identified modes.

To achieve that, the power fraction L_{cut}, which is *not* transmitted by the aperture, is without loss of generality split into two sums of transversal modes

$$L_{\text{cut}} = \sum_{(n,m) \in \{\substack{\text{Ident.}\\ \text{modes}}\}} C_{nm} (1 - T_{nm}) + \sum_{(w,v) \in \{\substack{\text{not ident.}\\ \text{modes}}\}}^{\mathcal{N}} C_{wv} (1 - T_{wv}) \quad . \tag{2.39}$$

Here the C_{ba} and the T_{ba} are the transverse mode coefficients as introduced in 2.1.2 and the power transmission coefficients through the aperture, respectively, of any transversal mode with mode indices (b,a). The parameter $\mathcal{N}$ denotes an effective number of not identified modes necessary to explain the integral power fraction L_{cut}, which is lost by transmission through the aperture. The transmission coefficients T_{ba} depend on the mode index pair (b,a), and on the beam parameters. Correspondingly, the value of $\mathcal{N}$ obviously depends on the specific transversal modes, which are incident on the aperture.

The transmission coefficients T_{ba} are explicitly calculated by

$$T_{ba} = \int_{0}^{d_{\text{ap}}/2} r \int_{0}^{2\pi} |V'_{ba}(z_{\text{ap}})|^2 \, d\phi \, dr$$

where V'_{ba} in this case denotes Laguerre Gauss modes (LG$_{ba}$), normalized such that T_{ba} results in unity for $d_{\text{ap}} \to \infty$. With the help of the Laguerre polynomials $L_{ba}(x)$ these normalized modes are given by [129]

$$V'_{ba}(z) = \sqrt{\frac{2\, b!}{2\pi\,(b+a)!}} \frac{1}{w(z)} \left(\sqrt{2}\frac{r}{w(z)}\right)^a L_{ba}\left(2\frac{r^2}{w(z)^2}\right) \exp\left(-\frac{r^2}{w(z)^2}\right)$$
$$\times \exp\left(i\,k\,\frac{r^2}{2\,R(z)}\right) \exp\left(i\,a\,\phi - i(2b+a+1)\psi(z)\right) \quad .$$

LG modes are better suited here because a circular aperture is used. The parameters appearing in this expression are explained in 2.1.2.

As one is interested here in a lower limit on η_{00}, one needs an upper limit on $\mathcal{N}$. This can be readily constructed when a worst-case scenario (concerning the

2.4. A 130 W CW single-frequency TEM$_{00}$ green laser source

sensitivity of the aperture method) is concerned. In any given situation the incident beam will consist of those transversal modes identified by the mode scan technique plus a certain amount of not identified modes, each of which can contain a maximum power fraction of γ_{max}. At this point the worst-case assumption is made, which claims that these not identified modes consist only of those modes, which are transmitted best through the aperture (further details are given below), and that all of these modes contain the maximum possible power fraction γ_{max}.

As the integral power fraction, which is *not* transmitted through the aperture is known to be L_{cut} from a measurement, a worst-case upper limit for the effective number of not identified higher-order transversal modes $\mathcal{N}$ can be obtained via a simple algorithm. It first calculates the T_{ba} of a sufficiently large sample of transversal modes. It then sorts these modes by their value of T_{ba}. And it finally finds the upper limit for $\mathcal{N}$ by summing up the fractions $\gamma_{\mathrm{max}}(1 - T_{\mathrm{wv}})$ of all not identified modes starting from highest T_{wv} until Eq. (2.39) is fulfilled. From the obtained upper limit on $\mathcal{N}$ the power fraction of the incident beam, which is not contained in the TEM$_{00}$ mode, is given by

$$(1 - \eta_{00}) \leq \frac{1}{P_{\mathrm{inc}}} P^{(\mathrm{ident.\ modes})} + \mathcal{N} \gamma_{\mathrm{max}} \quad . \tag{2.40}$$

The first ratio in this expression is known from the result of the spectrum mode of the mode analyzer.

In this procedure it was assumed, that the power fraction contained in all identified modes is not significantly reduced by transmission through the aperture, which is easily checked for with the result of the same algorithm just described.

2.4.7. Results and discussion of the SHG experiment

Measured conversion performance

The external conversion performance and the obtained harmonic power from the LBO SHG resonator are shown in Fig. 2.23. Each data point for the harmonic power is the average value of a continuous measurement of length 10 s. The error bars were derived from the calibration errors stated above. A maximum harmonic

2. High-power 532 nm single-frequency TEM$_{00}$ laser sources

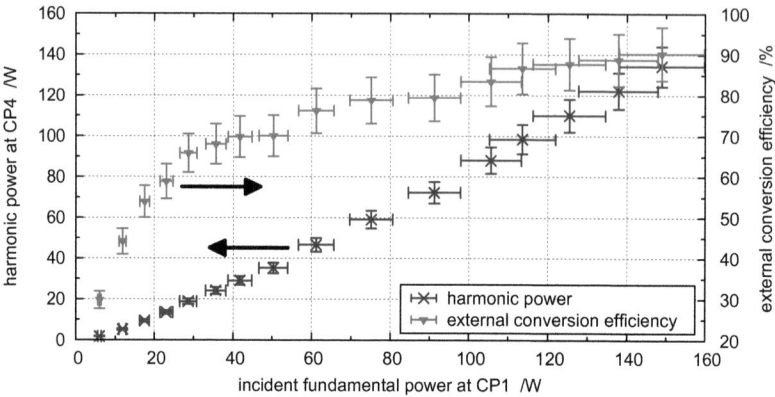

Fig. 2.23: Measured external conversion efficiency (right axis) and harmonic power at CP4 (left axis) of LBO SHG resonator.

output power of 134 W was measured at point CP4 when of fundamental power 149 W was measured at point CP1. To the best of the author's knowledge this is the highest power of CW single-frequency laser radiation in the green visible spectral region, which has been scientifically published so far. Similarly, it is the highest scientifically published power of green CW laser light generated by the effect of SHG in general, even taking multi-mode systems into account (for a compilation of reported values see 2.4.1).

The spectral linewidth of the harmonic light was measured to consist of a single spectral line narrower than 700 kHz on timescales of 1 s. This result was obtained from transmission of the harmonic light through the reference resonator of the mode analyzer. As the frequency conversion process does not influence the spectral distribution of narrow-linewidth fundamental light in lowest order, the spectral linewidth of harmonic light from SHG is in fact strongly expected to resemble that one of the primary laser. This primary laser was a non-planar ring oscillator (NPRO) with a spectral linewidth on the order of 100 Hz within a 25 ms measurement interval [35]. A representative measurement of the NPRO's fundamental frequency noise presented as a linear spectral density was reported in [81].

The relative peak-to-peak power fluctuations of the harmonic light amounted to 8 % within 100 s at maximum harmonic power. All in all the system was operated for more than 100 h at harmonic power levels above 110 W. No degradation of crystal or optics was measured within this time period. The system also reproduced the reported conversion performance after being switched off for a while. The control loop performance was robust with no obvious signs of coupling of the circulating or harmonic power to the stability of the frequency or temperature control loops.

Fig. 2.23 also presents the overall external conversion efficiency η_{SHG} of the SHG resonator as derived from the measured harmonic and incident powers (see Eq. (2.37)). An external conversion efficiency of more than 90 % was achieved at maximum harmonic power. Again, to the best of the author's knowledge this is the highest value reported in literature so far for harmonic powers above 150 mW. For harmonic powers below 150 mW a slightly higher value of 92 % was obtained from an etched zink doped PPLN ridge waveguide structure [141, 142]. However, from the discussion in 2.2.4 and 2.4.1 it is strongly expected by the author of this thesis, that this value would degrade considerably when scaling the harmonic power to multiple tens of watts.

Modelling the harmonic power

To further investigate the performance of the LBO SHG resonator presented here, the experimental data for the harmonic power was compared with the prediction of a model, which does not account for thermal dephasing processes in the crystal and which was described in 2.2.5. The comparison is shown in Fig. 2.24 for three different amounts of mode mismatch between incident beam and resonator eigenmode. All other parameters were the same and given essentially by Tbl. 2.2 with the exception of the measured values of $T_{\text{in}} = 25.5\%$ and $A_{\text{p}} = 1.4\%$. The latter value was obtained from a measurement of the power buildup when the phase matching temperature of the crystal was strongly detuned.

2. High-power 532 nm single-frequency TEM$_{00}$ laser sources

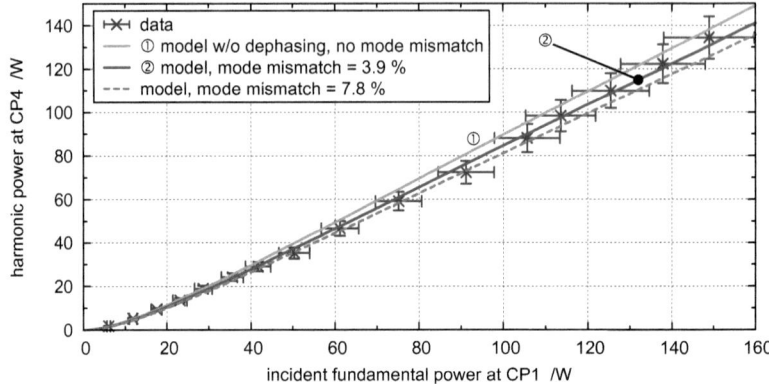

Fig. 2.24: Harmonic power at CP4 of resonant LBO SHG stage compared with three predictions of a model for different amounts of mode mismatch between incident beam and resonator eigenmode. The model did not account for thermal dephasing effects inside the crystal.

The reason for the consideration of the three distinct amounts of mode mismatch is explained in the following. In the experiment described here the mode mismatch was determined from a measurement of the fundamental power fraction, which was reflected from the SHG resonator in a nearly impedance matched situation. A measurement of this power fraction is shown in Fig. 2.25.

Unfortunately, modelling of this measurement was not clear without ambiguity. While the data at lowest and highest harmonic powers fit best to the simulation, if a mode mismatch of 4.4 % was assumed (which is not shown in Fig. 2.25), the data in between these regions fit best to the simulation, if no mode mismatch at all was assumed. The reason for this discrepancy is expected to be caused on the one hand by a varying mode mismatch from thermal lensing in the SHG resonator. On the other hand a completely vanishing mode mismatch (as pretended then by the incident power range between 20 W and 80 W) is rather unlikely, due to the slightly

2.4. A 130 W CW single-frequency TEM$_{00}$ green laser source

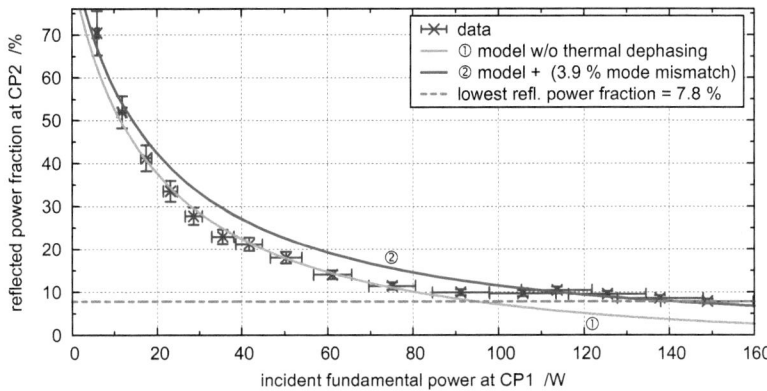

Fig. 2.25: Fundamental power fraction, which was reflected from the LBO SHG resonator compared with two predictions of a model (one for 3.9 % mode mismatch and one for no mode mismatch at all), which assumed no thermal dephasing effects inside the crystal. The horizontal broken line is a guide for the eye and indicates the lowest measured reflected fundamental power fraction of 7.8 %.

elliptic coupling waist, for example.

Therefore one also has to take into account varying contributions to the impedance mismatch from other sources than from conversion losses. In principle, there might have been two contributions to this. One would be caused by the effect of GRIIRA in the LBO crystal. The other would originate from changes of the reflectivity of the input coupler or the coatings on the crystal facets due to heating from optical absorption. In order to avoid a vanishing mode mismatch by the assumption of such a varying impedance mismatch contribution to the measurement of Fig. 2.25, a change of the round trip losses by at least $1 - 2$ % would have been necessary. However, this effect appears to be too strong to be caused by GRIIRA because it would have led to considerable thermal dephasing. This in turn would have led to a

2. High-power 532 nm single-frequency TEM$_{00}$ laser sources

significantly reduced harmonic power compared to the applied model, which is *not* observable in Fig. 2.24.

Summarizing these thoughts, the inconsistencies in the measurement shown in Fig. 2.25 should have been caused by two effects, one of which was thermal lensing in the SHG resonator. The other effect is assumed here to have been caused by reflectivity changes of the input coupler or coatings on the crystal facets by heating from optical absorption. Such effects cannot be accounted for by the model used to simulate the measurement of the reflected power fraction. Therefore one could derive from this measurement only an upper limit on the mismatch between incident mode and resonator eigenmode. This upper limit was conservatively taken to be the lowest reflected power fraction measured for any incident power, which amounted to 7.8 %. In turn, a conservative lower limit is obtained from the assumption of an indeed vanishing mode mismatch. These limits are shown in Fig. 2.25 and they were also used to model the result for the harmonic power in Fig. 2.24. The region of harmonic power levels allowed by these limits nicely agree with the measured values within their error bars. As guide for the eye also the average value of these limits was used to model the harmonic power.

One should note here, that this agreement between the data and a model, which does not account for thermal dephasing effects in the crystal, clearly demonstrates that such effects did not have any significant influence on the SHG resonator performance up to maximum power levels! This is indeed remarkable, as these maximum power levels were as high as 640 W for the circulating fundamental power (see below) and 134 W for the harmonic power. These power levels corresponded to peak intensities in the crystal of up to 10 $\frac{\text{MW}}{\text{cm}^2}$ for the fundamental beam and 4 $\frac{\text{MW}}{\text{cm}^2}$ for the harmonic beam.

Limiting factors for the conversion efficiency

As was explained in the context of Eq. (2.35), the external conversion efficiency η_{SHG} of a realistic experiment will be limited by three parameters. These are the ratio of passive and conversion round trip losses $r = A_\text{p}/A_\text{SHG}$, the fraction of incident power

2.4. A 130 W CW single-frequency TEM$_{00}$ green laser source

in the resonator's TEM$_{00}$ mode η_{00}, and the transmission T_{dc} of the dichroic mirror. The values for these parameters at maximum harmonic power, which were derived in the previous paragraphs, are listed in Tbl. 2.3. The value for A_{SHG} was obtained from the simulation of the $(1 - \eta_{00}) \approx 3.9\%$ case in Fig. 2.24.

parameter	value
1-T_{dc}	2.5 %
$(1-\eta_{00})$	$\approx 3.9\%$
$g(r = 1.4\%/21.9\%)$	6.0 %
sum	$\approx 12.4\%$

Table 2.3: List of parameters, which limit the external conversion efficiency of the LBO SHG resonator.

Summation over all contributions results here in a value of $\mathcal{L}_{SHG} = 12.5\%$. In analogy to the concept of external conversion efficiency one might think of $\mathcal{L}_{SHG}$ as an external conversion loss. Its value as derived from the summation above is only slightly higher than what is expected from the measured maximum conversion efficiency of $\eta_{SHG} = 1 - \mathcal{L}_{SHG} = 90\%$. The discrepancy might arise from the only roughly estimated value of η_{00} or because the dependence of $g(r)$ on r has flattened due to depletion of the fundamental light. In the form given, passive losses of the resonator had the biggest impact on η_{SHG}.

Circulating power and linear fundamental absorption

To complete the performance analysis, also the fundamental power circulating inside the SHG resonator at CP3 was measured and modelled. The comparison of the data with the simulation results is presented in Fig. 2.26, where the same limits on the amount of mode mismatch were used as in the figures above.

For the case of an external-cavity doubling scheme the circulating power level is more sensitive to the onset of thermal dephasing processes than the harmonic power (see discussion in the context of Fig. 2.14 in the previous section). In this respect

2. High-power 532 nm single-frequency TEM$_{00}$ laser sources

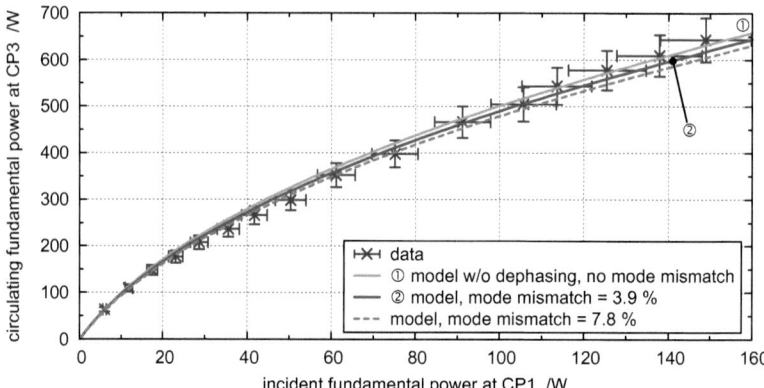

Fig. 2.26: Circulating fundamental power inside LBO SHG resonator compared with three predictions of a model for amounts of mode mismatch of 7.8 %, 3.9 % and vanishing mode mismatch. The model does not account for thermal dephasing effects inside the crystal.

the weak apparent rise of the measured data at maximum circulating powers above the simulation *might* be an indication for the onset of thermal dephasing processes inside the crystal.

The circulating power level reached high values of up to 640 W. In the context of Fig. 2.25, it was assumed that thermal lensing inside the optical resonator might vary the mismatch between incident and circulating mode. To analyze, if linear absorption at the fundamental wavelength could lead to considerable thermal lensing in the crystal, efforts were made to obtain the linear absorption coefficient $\alpha_{l,f}$ from an independent thermal lensing measurement. This was conducted in the same way as described in the context of Figs. 2.11 and 2.12 in the previous section. Its result is shown in Fig. 2.27 [24]. The experiment itself and its evaluation were again not done by the author of this thesis.

2.4. A 130 W CW single-frequency TEM$_{00}$ green laser source

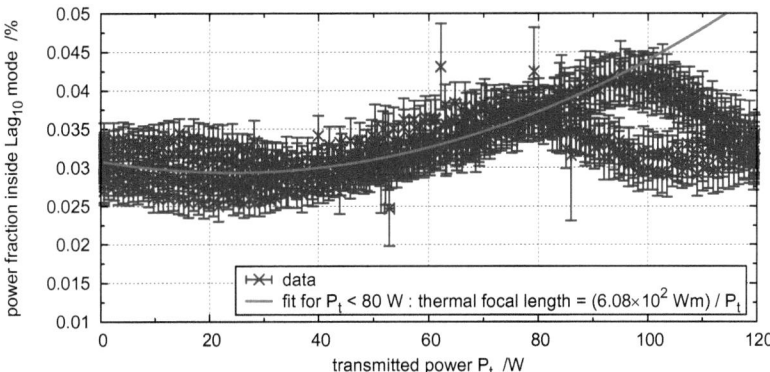

Fig. 2.27: Measurement of the power fraction of a TEM$_{00}$ laser beam at 1064 nm, which was transferred into the LG$_{10}$ mode by transmission through a LBO crystal far from its phase matching temperature. The crystal was mounted in the same way as for the resonant LBO stage. The beam was focussed to a waist size of 76 μm located 40 mm off the crystal center.

Obviously the result in this case was dominated by systematic errors of the experiment, as the drop for highest powers implies. To derive at least an upper limit on the linear fundamental absorption coefficient the range of dropping mode content was ignored, when fitting the result. In the same way as described in the previous section the fit result was compared with the simple model for a thermal lens given by Eq. (2.38). The comparison is shown in Fig. 2.28. This method resulted in a surprisingly low upper limit for the linear absorption coefficient of the LBO crystal for 1064 nm light of $\alpha_{l,f} \leq 4.2\times10^{-4}\,\frac{1}{\mathrm{m}}$. This value is much smaller than those reported in literature (see 2.2.4). It is indeed comparable in size to the absorption coefficients of high-quality Suprasil substrates, which are on the order of $\alpha_{l,f} \approx 2\times10^{-4}\,\frac{1}{\mathrm{m}}$ [89].

Such a low value for the linear absorption coefficient for 1064 nm radiation implies that at maximum circulating power inside the LBO SHG resonator the crystal was

2. High-power 532 nm single-frequency TEM$_{00}$ laser sources

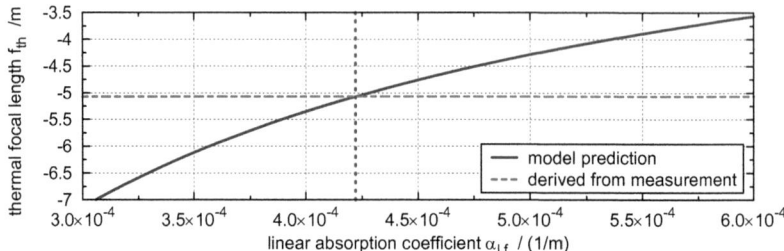

Fig. 2.28: Comparison of the prediction of a model with the measured upper limit for the thermal lens in LBO from absorption of 1064 nm light. The model matches the measured upper limit for an absorption coefficient of $\alpha_{l,f} = 4.2\times10^{-4}\,\frac{1}{\mathrm{m}}$, indicated by the vertical dotted gray line.

heated with a power of only 13 mW. Therefore, it should not have been a source of thermal lensing. However, thermal lensing could still have taken place in the resonator if considerable heating of the resonator mirror facets occurred (see [152] for a model of this).

The very low dissipation inside the crystal, which was found by this result as well as by the results from the previous paragraphs, fitted well to the easy experimental handling and robust control loop performance of the SHG resonator. Neither the frequency control loop nor that one of the oven seemed to be influenced by larger fluctuations of the circulating or harmonic power. This was very different from the experiences with the resonant PPKTP SHG stage described in the previous section.

Temperature tuning curve and linear harmonic absorption

Many publications, which report on efficiency deviations of SHG devices from theoretical prediction under high power operation, also report on distortions and shifts of the temperature acceptance curves ideally given by Eqns. (2.25) and (2.26) (see 2.4.1 for a compilation of literature). Hence still another way to search for the possible

2.4. A 130 W CW single-frequency TEM$_{00}$ green laser source

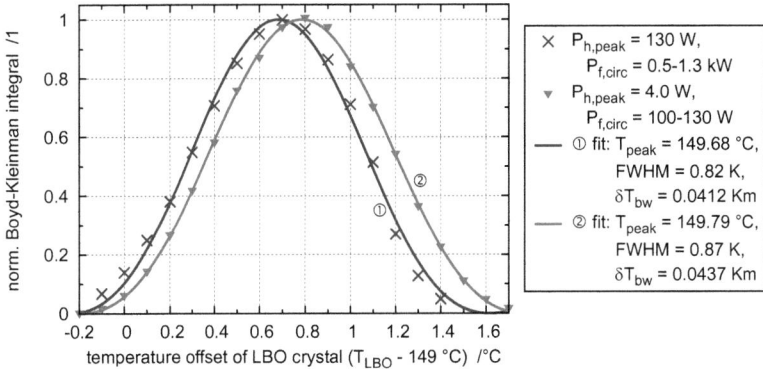

Fig. 2.29: Temperature tuning curves of LBO crystal for low and maximum harmonic peak power while located inside SHG resonator. From the measured values of P_{circ}(CP3) and P_h(CP4) the value of the Boyd-Kleinman integral $\hat{h}$ was derived by application of Eq. (2.28), which was solved for $\hat{h}$.

onset of such effects is to measure this temperature acceptance curves for high and low power levels. This was done and the result is presented in Fig. 2.29 together with fits of the above mentioned equations to the data.

The obtained values for the temperature acceptance bandwidth δT_bw agree well with the values given in literature, for instance 4.2×10^{-2} Km from [85]. Obviously only small changes of the temperature acceptance curves were measured when increasing the harmonic peak power by more than a factor of 30 and the circulating fundamental power depending on the temperature by up to a factor of 10. The range of values for the fundamental power arose because the crystal was located inside the SHG resonator, whose power buildup varied depending on the conversion losses. This was necessary to measure the temperature tuning curve at maximum harmonic power.

2. High-power 532 nm single-frequency TEM$_{00}$ laser sources

Apart from a tiny inclination of the curve for high harmonic power, only a small shift of the optimum phase matching temperature by 0.11 K was observed. In the following it is assumed that this temperature offset was caused by absorption from the harmonic and fundamental beam at roughly equal amounts. This absorption then created a temperature gradient from the crystal's center to its border, which can be expressed with the same approximations made in the development of the dephasing model around Eq. (2.36) by

$$\Delta T = \frac{P_{\text{heat}}}{4\pi\, L_k\, \kappa_{\text{th}}} \left[0.577 + \ln\left(2\left(\frac{r_0}{0.85\, w_{0,f}}\right)^2 \right) \right] \quad . \tag{2.41}$$

In this equation the crystal is again assumed to be a rod. In contrast to the dephasing model, here the heat source radius is given by the average of fundamental and harmonic waist, and it is compared to the radius of the crystal r_0. The latter point is necessary because in this kind of measurement the phase matching temperature is intended to be *not* optimized for each data point. From the equation the measured shift of the temperature tuning curve is obtained for a heating power of $P_{\text{heat}} = 31$ mW with a thermal conductivity of LBO of $\kappa_{\text{th}} = 3.5\,\frac{\text{W}}{\text{Km}}$ and a radius of the LBO crystal of $r_0 = 1.5$ mm. On the one hand this value justifies the assumption made above, that the temperature profile in the crystal was created by roughly equal amounts of heating from both wavelengths. On the other hand the linear absorption coefficient for the harmonic light in the LBO crystal can be estimated from the values for P_{heat} and $\alpha_{l,f}$ to be $\alpha_{l,h} \approx 2.6 \times 10^{-3}\,\frac{1}{\text{m}}$. This value is also rather tiny but nonetheless roughly matches the value reported by [92].

With estimations for $\alpha_{l,f}$ as well as $\alpha_{l,h}$ at hand one can employ the thermal dephasing model again to simulate the harmonic power expected from the resonant LBO SHG stage with thermal dephasing taking place in the crystal. The result is, however, identical to the prediction of the model neglecting thermal dephasing within less than 0.1 % of all measured harmonic power levels. This strengthens the result obtained above, that in spite of the extraordinary high powers involved, thermal dephasing was of no importance for this SHG resonator.

2.4. A 130 W CW single-frequency TEM$_{00}$ green laser source

Comparison with resonant PPKTP SHG stage

Once the conversion performance and absorption coefficients are known, another interesting question is, in which range of input powers an impedance matched LBO SHG stage as described in this section would be more efficient than the PPKTP SHG stage, which was described in the previous section. This question is answered by Fig. 2.30. For the simulations in this figure it was assumed that the transmission coefficient of the dichroic mirror $T_{dc} = 97.5\%$ and the passive round trip losses $A_p = 1.4\%$ were the same for both devices, and that each device was impedance matched for each datapoint anew. All other parameters had those values reported at the appropriate places in the text.

Obviously the LBO SHG stage is more efficient than the settled PPKTP SHG stage for all incident power levels above already 2 W. This power limit is increased to 8 W for the case of the PPKTP SHG stage in its initial state before settling of its harmonic power due to gray tracking occurred.

In any case the LBO SHG stage described in this section seems to be the better choice for the conversion of even intermediate power infrared lasers.

Long-term stability

To demonstrate the long-term power stability of the LBO SHG resonator it was equipped with automatic lock acquisition electronics and its incident and harmonic power were monitored for 48 h after a warming period of 13 h. During this warming period the SHG resonator had to be realigned a few times to keep the harmonic power close to maximum. During the 48 h of the measurement the SHG resonator was realigned only once about half an hour before the measurement was ended. The time series of the harmonic power is presented in Fig. 2.31. The incident power level was constant over the whole time. In this measurement the harmonic power was only 112 W because the primary infrared laser did not emit its full power. The duty cycle of the SHG resonator was 99 % and the longest continuous duration without loss of the stabilized state lasted 6 h. After each loss of the locked state it was acquired again automatically within $1 - 2$ s. The time periods with frequent losses

2. High-power 532 nm single-frequency TEM$_{00}$ laser sources

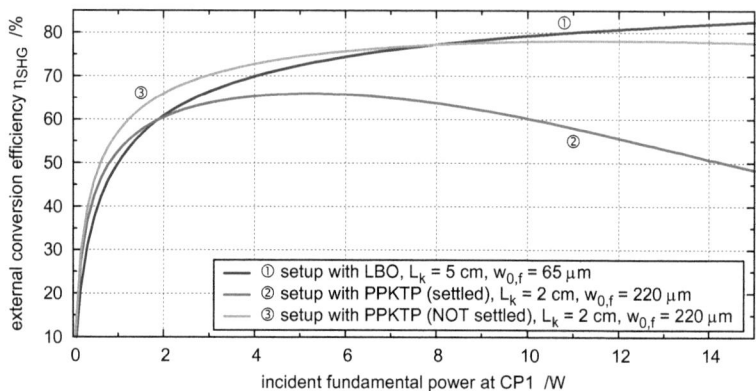

Fig. 2.30: Comparison of external conversion efficiency of resonant LBO stage (as described in this section) with resonant PPKTP stage (as described in the previous section). The values for $T_{\text{dc}} = 97.5\,\%$ and $A_{\text{p}} = 1.4\,\%$ were assumed to be equal for both devices. The input coupler transmission coefficient was chosen for each device and each incident power level such that impedance matching was realized. All other parameters had the values given at the appropriate places in the text.

of the locked state (visible as significant accumulations of power reductions) can be avoided in the future by increasing the dynamic range of the fast frequency actuator of the resonator by approximately 50 % (e.g. by employing a longer piezo or a higher maximum voltage).

The peak-to-peak power variations over the full measurement duration (excluding losses of the locked state) amounted to 10 %. The gradual weak reduction of the harmonic power was due to slowly drifting alignment of the SHG resonator eigenmode or the incident mode relative to each other. To demonstrate this, the setup was manually realigned at the end of the measurement and the data acquisition was continued for another 30 min. By this manual realignment essentially the same

2.4. A 130 W CW single-frequency TEM$_{00}$ green laser source

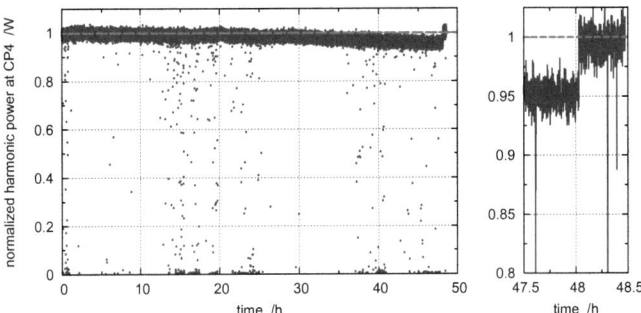

Fig. 2.31: Measurement of the harmonic power stability of the LBO SHG resonator on long time scales at a harmonic power level of 112 W (left graph), and a zoom of the time period when a manual realignment was made about half an hour prior to the end of the measurement (right graph). The red broken line is a guide for the eye and indicates the power level at the start of the measurement.

power level could be obtained as was emitted at the beginning of the measurement. The right hand side of Fig. 2.31 shows a zoom to the point in time when the realignment was made to demonstrate this more clearly. To get rid of this problem one might either construct a more rigid resonator or implement an auto-alignment system.

In sum the SHG resonator was operated for more than 100 h at harmonic powers above 110 W without observable performance loss due to degradation of the crystal or optics.

Transverse mode analysis

As the LBO SHG resonator was designed as a metrology laser source, its beam quality (i.e. the harmonic power fraction contained in higher-order transversal modes) was analyzed, too. The results from the various operational modes of the mode

2. High-power 532 nm single-frequency TEM$_{00}$ laser sources

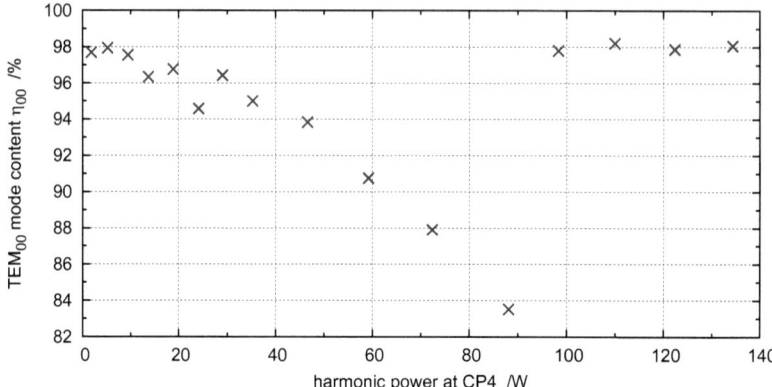

Fig. 2.32: The mode analyzer in spectrum mode was used for each data point of Fig. 2.23 to determine an upper limit on the fraction of the harmonic power η_{00}, which was contained inside the TEM$_{00}$ mode. The alignment to the mode analyzer was optimized for each data point but the positions of the mode shaping lenses were optimized only once at a harmonic power of 90 W.

analyzer are presented in the following.

Spectrum mode. For each data point in Fig. 2.23 the mode analyzer was used in its *spectrum mode* to determine the transversal mode structure. The experimental setup, theory and data evaluation process of this operational mode was already described in 2.4.4 and 2.4.5. The result of the spectrum mode is presented in Fig. 2.32 where the harmonic power fraction η_{00} contained within the TEM$_{00}$ mode is displayed as a function of the harmonic power available at point CP4.

In a first step the value at low harmonic powers was optimized by compensation of a small astigmatism of the harmonic beam that originated from its transmission through the curved and angled dichroic resonator mirror and that could be observed by a non-vanishing amount of power in the TEM_{02} mode. By tilting the second mode matching lens of the mode analyzer horizontally by approximately 7 deg this astigmatism could be reduced leading to $\eta_{00} \approx 98\%$ of the power being emitted into the fundamental transversal mode at low powers.

When the harmonic power was increased this fraction reduced, although the alignment to the mode analyzer cavity was optimized for each data point. Especially the power fraction inside the TEM_{02} and TEM_{20} modes increased. Their resonance frequencies were degenerate and formed a single resonance for the LG_{10} mode. The increase of the fractional power in these modes hints on a change of the harmonic mode shape, which most likely originated from thermal lensing. From the investigation above it can be concluded that this thermal lens was *not* located inside the LBO crystal. Suspects for the origin of this thermal lensing are every resonator mirror, which could have deformed by heating (see [152]), and the two mirrors, which directed the high harmonic power to the dump (see Fig. 2.21). At a harmonic power of 90 W the positions of the lenses, which matched the incident mode to the eigenmode of the mode analyzer cavity, were slightly changed to compensate for the thermal lensing and to maximize η_{00} again. This step restored the high fundamental mode content of $\eta_{00} = 98\%$, which surprisingly did not decrease any more for higher harmonic powers.

Transmission mode. As was described in 2.4.5 the result of the spectrum mode of the mode analyzer might have been an upper bound for η_{00} only. For the highest harmonic power also a strict lower bound should be obtained and thus the mode analyzer was also used in its *transmission mode*. The lower bound, which was obtained from this measurement was $\eta_{00} \geq 91\%$.

Aperture mode. To reduce the difference between the two limits for η_{00} obtained so far, the mode analyzer was also operated in its *aperture mode* as described in 2.4.6. The chosen diameter and position of the aperture described above resulted in an integral transmission of 99.6 % of the incident beam through the aperture and thus in a value of $L_{\text{cut}} = 0.4\%$. The algorithm described above had to sum up the calculated transmitted powers of $\mathcal{N} = 28$ hypothetical higher-order modes with mode index sums up to 10. All these transversal modes were correspondingly assumed to be part of the laser beam, but at the same time were *not* identified by the spectrum mode of the mode analyzer. Those modes, which were identified by the spectrum mode, were inspected visually via a CCD camera to determine their mode index sums. The corresponding mode scan was already shown above in Fig. 2.22. These modes were not included into the summation algorithm just mentioned.

The measurement in the aperture mode resulted in a maximum fraction of harmonic power within possibly not-resolved higher-order modes of 1.4 %. The power fraction in higher-order modes already identified in the spectrum mode amounted to 1.9 % and only 0.005 % of this already identified higher-order mode content would have been cut off by the aperture. Thus the aperture mode determined a lower bound for the power fraction within the TEM_{00} mode of 96.7 %. All in all this limits the power fraction within higher-order transversal modes $(1 - \eta_{00})$ at a harmonic output power of 134 W to the remarkably low range of values of $1.9\,\% \leq (1 - \eta_{00}) \leq 3.3\,\%$. This in turn resulted in at least 130 W of laser radiation at 532 nm in the TEM_{00} mode!

2.5. Summary and outlook

The availability of a metrology laser source at a wavelength of 532 nm, which emits more than 100 W, would help to increase sensitivities or efficiencies in many fields of research. Often single-frequency emission and a high beam quality are also necessary, e.g. in the search for gravitational waves. While such laser sources could already be constructed at other wavelengths, the green visible spectral region proofed hard to cover. In this chapter the solution was found by application of the nonlinear effect of second harmonic generation.

In Section 2.1 and Section 2.2 the basics of this nonlinear effect were reviewed and the limitations and problems were explained, which prevented the realization of a high-power green laser source up to now. As high intensities and absorption inside the nonlinear crystal were found to be a main concern, a simple model was developed, which showed the way to reduce the intensities. It was further extended to incorporate thermal dephasing effects from optical absorption inside the crystal to some extend.

To test the potential of an SHG resonator with by far reduced intensities inside the crystal, a resonant SHG stage based on PPKTP was set up and characterized precisely, which was presented in Section 2.3. This experiment resulted in a long-term stable harmonic power of 5.3 W, which is, to the best of the author's knowledge, the highest long-term stable harmonic power obtained from PPKTP at a wavelength of 532 nm. Nonetheless the performance was still not optimal and considerable efforts were made to identify the limiting effect. This was found to be thermal dephasing from absorption of the harmonic light, which could be modelled consistently.

Finally in Section 2.4 a very high-power infrared laser was converted to 532 nm. Based on the experiences of the previous section here another strategy was pursued, which relied on LBO as nonlinear material. This material offers a rather low nonlinearity, but at the same time it was known to show low absorption and little sensitivity to high intensities in any other way. The result was remarkably good, as a huge harmonic power of 134 W could be generated with a very high external conversion efficiency of 90 %. To the best of the author's knowledge the first value is the highest scientifically published so far for the generation of CW single-frequency green laser radiation and also the highest for SHG of CW green light in general. The external conversion efficiency is the highest reported value for harmonic powers above 150 mW. The SHG resonator was easy to handle and the harmonic power level was long-term stable. Comparison with a model revealed that thermal dephasing effects in the crystal were not at all an issue even at maximum harmonic power.

Additionally, a new technique was developed and presented in Section 2.4, which allows to reduce the uncertainty in the determination of the higher-order mode content of a laser beam. It was applied as one technique among others to characterize the beam quality of the LBO SHG resonator. These measurements found that a fraction of $97-98\,\%$ of the maximum harmonic power was contained in the TEM_{00} mode, which makes the overall device *a 130 W CW single-frequency TEM_{00} long-term stable 532 nm metrology laser source*!

2. High-power 532 nm single-frequency TEM$_{00}$ laser sources

The maximum harmonic output power achieved in this chapter appeared to be limited almost exclusively by the available fundamental power. This was derived from the fact that the performance of the LBO SHG resonator could be well modelled without accounting for thermal dephasing processes in the crystal. Only the measured fundamental circulating power *might* have deviated slightly from this model.

If the harmonic power level of this resonator design would be doubled the laser-induced damage thresholds of the coatings applied in this experiment are reached. But this can then be overcome either by utilization of the method of fundamental waist enlargement as presented in this chapter and applied in the PPKTP experiment. Or it can be overcome simply by the employment of higher-quality coatings, e.g. produced by the technique of ion-beam sputtering (IBS).

The external conversion efficiency of the current setup is expected to rise with increasing incident power and decreasing passive round trip losses. This will work until either thermal dephasing processes become observable, or until it gets limited at about 93.5 % mainly by imperfect matching of the incident mode and eigenmode of the SHG resonator. Further improvement of this matching should be possible to some extend, e.g. by making the incident beam slightly elliptic to match it better to the likewise slightly elliptic coupling waist of the SHG resonator.

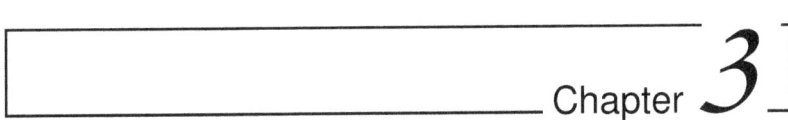

Chapter 3

ALPS I project - Particle physics with high-power green light

Thanks to the impressively precise results of the WMAP satellite mission, physicists now have compelling proof that only about 4.6% of the universe we live in consists of those types of matter that we know about and understand [70]. The physics of all the rest, which is usually denoted as dark matter and dark energy, are completely unknown, which means it does not fit into the generally accepted SM.

Since long particle physicists have developed and extended this model in order to explain as much observable phenomena as possible. As it is the highest mark and justification of physics to arrive at a single unified and consistent model of this universe, and as some of those just mentioned SM extensions contradict each other, it clearly has to be the aim of physics to find experiments, which are able to approve some of these extensions and falsify the others.

The scientific field of experimental particle physics is usually associated with the measurement technique of particle accelerators like the Large Hadron Collider (LHC) at CERN [31]. These huge and costly devices are used to test predictions of theoret-

3. ALPS I project - Particle physics with high-power green light

ical particle physics as well as to look for new currently not understood processes. Apart from this specific experimental technique also other possibilities exist to probe new particle physics, one of these being the search for the effect of LSW [4]. Here the regenerated photons from an as strong as possible laser beam are searched for inside a low background environment, which is optically shielded from the beam. Various models exist that predict the conversion of a small fraction of the photons into various hypothetical particles, which (similar to neutrinos) hardly interact with matter and thus will make their way through the shielding into the low background environment, where a similar fraction would convert back to photons. Some models require strong magnetic fields for this effect to take place.

Several large-scale LSW experiments based on pulsed laser sources have already been performed in the world, arriving at an impressively high resolution for the arrival of single photons and at correspondingly strict constraints on the particle physics models tested (e.g. [33]). The sensitivities of these experiments were limited by the average power of available pulsed lasers. Utilization of an external cavity to resonantly enhance the light in the production region instead inside the laser with its absorbing inner components turned out to be a crucial factor for further sensitivity enhancements.

Hence in the context of this thesis the optical injection stage and a nearly 9 m long external optical resonator comprising the particle production region of an existing large-scale LSW experiment were designed, implemented, characterized and maintained. This LSW experiment was dubbed 'Any Light Particle Search' (ALPS) experiment, located at DESY, Hamburg, and it was set up around an approximately 15 m long cryogenically cooled accelerator magnet. In a first step the compatibility of a long optical resonator and a large-scale LSW experiment was demonstrated. In the next step the experiment's sensitivity was optimized, mainly by implementation of a high-power resonant SHG stage and by the improvement of the production resonator. The superiority of the applied concept was proven by obtaining the by far most stringent constraints from laboratory experiments published so far on the particle physics models tested in a broad mass region.

In Sec. 3.1 of this chapter the Standard Model of particle physics, the basics of

LSW experiments and some phenomena and observations, which do not fit into the Standard Model, are described. After that Sec. 3.2 introduces those kinds of hypothetical particles, which were searched for. It also lists various cases of physics were the existence of those particles would significantly improve the understanding of observations or measurements. Sec. 3.3 lists briefly the state of the art of other LSW experiments in the world when the work at ALPS started, whereupon Sec. 3.4 explicitly explains and discusses the design, experimental setup and results of the compatibility demonstration experiments, dubbed ALPS I (phase 1). Finally, Sec. 3.5 does the same for the final ALPS I (phase 2) experiments, which achieved the most stringent laboratory constraints on the particle models so far.

The most important results of this chapter have already been published in [44] and in [45]. The impact of this latter publication for the field of research was stressed by the publication of a referencing note by *Nature* in their *Research Highlights* section [57]. Additional experimental details on the optical parts have already been published in [97].

3.1. Particle physics with laser light

3.1.1. The standard model of particle physics

All physical phenomena, which have been observed in the laboratory (or at particle colliders) until today can be explained by the interaction of only four forces with matter [38, 113]. These forces are

- the electromagnetic force being responsible for example for the existence of light and the formation of atoms and molecules,

- the weak force, which is relevant mainly in the context of decay processes like the beta decay,

- the strong force, which binds the constituents of the nucleons together and whose remnants are observable as the forces, which keep the atomic nuclei stable

3. ALPS I project - Particle physics with high-power green light

class	particle	flavour	electric charge	rest mass	spin	color charge
leptons	ν_e	1	0	< 0.7 eV/c^2	1/2	--
	ν_μ	2	0	< 0.7 eV/c^2	1/2	--
	ν_τ	3	0	< 0.7 eV/c^2	1/2	--
	e$^-$	1	-1 e	510 keV/c^2	1/2	--
	μ^-	2	-1 e	110 MeV/c^2	1/2	--
	τ^-	3	-1 e	1.8 GeV/c^2	1/2	--
quarks (bound)	up	1	+2/3 e	2.4 MeV/c^2	1/2	r/g/b
	charm	2	+2/3 e	1.3 GeV/c^2	1/2	r/g/b
	top	3	+2/3 e	170 GeV/c^2	1/2	r/g/b
	down	1	-1/3 e	4.8 MeV/c^2	1/2	r/g/b
	strange	2	-1/3 e	104 MeV/c^2	1/2	r/g/b
	bottom	3	-1/3 e	4.2 GeV/c^2	1/2	r/g/b

class	particle	flavour	electric charge	rest mass	spin	color charge
anti-leptons			one anti-lepton with inverted electric charge for each lepton			
anti-quarks			one anti-quark with inverted electric charge for each quark			

class	interaction	couples to	particles	rest mass	spin
bosons	electro-magnetic	electric charge	1 photon	0	1
	weak	weak charge	3: Z^0, W$^{+/-}$	91 GeV/c^2, 80 GeV/c^2	1
	strong	color charge	8 gluons	0	1
	inertia	rest mass	1 Higgs (?)	> 114 GeV/c^2	0

Fig. 3.1: The standard model of particles physics (SM) consisting of the fermionic constituents of matter (top left), which each have a corresponding anti-particle (top right), and of the bosonic force mediating particles (bottom) [38, 113, 9].

- and finally the gravitational force, responsible for the attraction of masses.

As far as particle physics at laboratory energies is concerned the gravitational force is of little importance because it is extremely weak compared to the other three. Furthermore its incorporation into quantized particle physics models proved to be difficult due to gravity being merely a property of spacetime instead of a force and because it is unclear how to quantize the equations of motion of general relativity [119, 113].

However, the remaining three forces could be formulated as exchange of bosonic particles and merged with a set of fermionic particles describing matter and another special boson into a single model known as the SM (see Fig. 3.1).

The group of leptons consists of the very light (but not massless according to the existence of neutrino oscillations) neutrinos, which couple only to the weak force, and

3.1. Particle physics with laser light

the significantly heavier electron, muon and tau, which couple to the electromagnetic and weak force. The quarks couple to the electromagnetic and weak force and they additionally carry a so-called color charge, which is a special property the strong force couples to. They are the constituents of the nucleons and can also temporarily form other particles (e.g. pions), but they have never been observed in an unbound state. Because they are fermions, leptons and quarks all have a spin of 1/2 and the overall number of fermions in the universe is conserved. Finally, there exists an antiparticle for each lepton, whose properties are the same as those of the particle but with inverted charges. Thus there are 48 leptons in the standard model.

To represent the different interaction types there are eight massless bosons called gluons, which couple to the color charges and thus mediate the strong force, one massless boson called photon, which couples to the electric charge and via this mediates the electromagnetic force, and finally the three very heavy bosons $W^\pm$ (also electrically charged) and Z^0 (electrically neutral), which couple to the so-called weak hypercharge property of the fermions and represent the weak force. All the bosonic force representing particles are spin 1 particles. Thus there are 12 force mediating bosons in the SM.

All these particles have already been experimentally observed. But finally, there is still another particle included in the SM, which has not been observed to date. It is called Higgs boson and it is used to theoretically describe the effect of inertia of particles. Thus it couples to all particles with non-vanishing rest mass. It is expected to be rather heavy ($> 114\,\text{GeV}$) and one hopes to directly detect it experimentally for the first time at the LHC in Geneva [31].

Summarizing, the SM consists of 61 particles. The structure of the SM is that of a set of gauge theories of quantized fields representing interacting particles in a flat spacetime obeying special relativity. Today it is generally accepted in the particle physics community (and presumably beyond) and many of its predictions could already be confirmed experimentally.

3. ALPS I project - Particle physics with high-power green light

3.1.2. Virtual particles and Feynman diagrams

As in the classical world all interactions of *real* particles in the context of the SM have to fulfill energy and momentum conservation. Here the term *real* denotes particles, whose existence can be proven directly by their detection. However, being a quantized model, the SM also knows so-called *virtual* particles. These particles are a pictorial representation for quantum fluctuations and higher-order processes. As such they can evade energy conservation locally because they exist only for a limited amount of time Δt and due to Heisenberg's uncertainty principle can spontaneously possess an energy amount ΔE given by:

$$\Delta E \, \Delta t \leq \hbar \quad .$$

In order not to violate other conservation laws (e.g. charge), virtual particles are produced as a pair of a particle and its antiparticle, who will annihilate each other after the limited period of time Δt. Therefore this process is often also called a *loop*. Antiparticles have inverted charge and lepton number but the same mass and spin as the corresponding particle. Because rest mass is equivalent to energy, very heavy virtual particles generally have shorter lifetimes. The large rest mass of the bosons of the weak force is therefore the reason for the very short interaction range of this type of interaction (in fact the shortest of all forces) at low energies [113].

A typical application of virtual photons is to provide a theoretical description for stationary fields, which becomes necessary if, for example, the Coulomb interaction between two approaching electrons is to be described. The virtual photon carries momentum from one electron to the other and thus scattering can occur. Interactions between the different particles are described by so-called Feynman diagrams. Fig. 3.2 shows the lowest order Feynman diagram for the elastic scattering of two electrons. These diagrams are created in the center of mass reference frame of the depicted process with time increasing from left to right. Lines with open ends at the sides stand for in- and outbound real particles while lines without open ends further to the center depict virtual particles. Antiparticles are travelling into opposite time direction. Different particles are depicted by different line types and within quantum

3.1. Particle physics with laser light

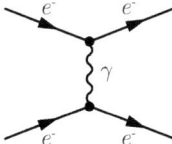

Fig. 3.2: Lowest order Feynman diagram depicting elastic scattering of two electrons.

field theory a mathematical expression can be assigned to each part of the diagram, which enables the calculation of the probability of this process to take place [113].

3.1.3. One-loop process magnetic vacuum birefringence

Another typical example for the occurrence of virtual particles is in the case of photon-photon interactions as for the effect of magnetic vacuum birefringence [60], which interestingly was already treated theoretically by Werner Heisenberg in 1936. Here a beam of light is shone through a strong and extended static magnetic field, which turns the vacuum into a dispersive medium. Fig. 3.3 shows the Feynman diagram of this process. Although the photon energy of visible light is orders of magnitude below the rest masses of an electron-positron pair, these particles can still be created as virtual particles for short time scales. The production probability is suppressed by the lack of energy. Such a process is denoted as a one-loop process and is mathematically treated as a process of higher order because in lowest order the photon would just pass the magnetic field without interaction. As electrically charged particles they will interact with the virtual photons of the magnetic field. Thus the vacuum effectively appears to be charged, correspondingly gets magnetized and light propagation through it has to be explained in analogy to light propagation through a medium in a magnetic field. This means the vacuum acquires a refractive index depending on the orientation between light field and magnetic field (see 2.1) [61, 3].

This effect is understood completely within the SM. However, is predicted to be very tiny. One expects a difference between the refractive indices of two perpendic-

3. ALPS I project - Particle physics with high-power green light

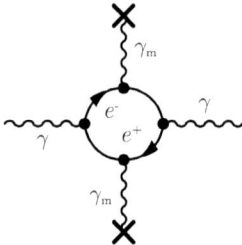

Fig. 3.3: Feynman diagram describing the SM effect of magnetic vacuum birefringence. The crosses denote the static magnetic field.

ular optical polarizations of [61]

$$\Delta n \approx 4\times 10^{-24} \times \left(\frac{B}{1\,\text{T}}\right)^2 \quad .$$

Present day experiments like the BMV experiment (see 3.3) tries to measure this effect but have to be improved by approximately two to three orders of magnitude [87, 16]. Interestingly, the gravitational-wave detector GEO 600 might in principle be able to measure this effect. As a very basic idea one of its arms could be equipped with a 100 m long and 0.3 T strong magnetic field. A magnetic field of this strength might be obtainable from several standard room-temperature coils with iron core, enabling the experimenter to turn the magnetic field on and off at approximately 500 Hz to work at the detector's best displacement sensitivity. Integration of the resulting periodic signal at the detector's output for about 2×365 days should then allow to isolate it from the background noise with a signal to noise ratio of unity (for GEO 600 and its sensitivity see [62] and references therein).

3.1.4. Particle accelerators and LSW experiments

In the scope of this thesis virtual particles are mainly important in still another context. As was explained above, two particles of the SM can in first order only interact with each other, if the first one carries properties (electric, weak or color

3.1. Particle physics with laser light

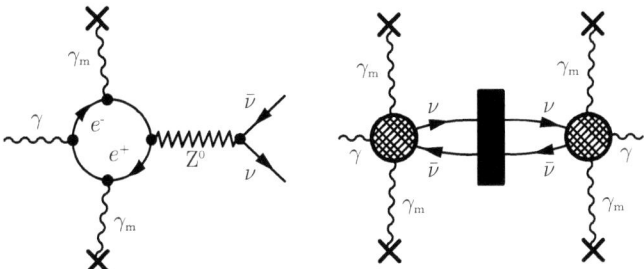

Fig. 3.4: Feynman diagrams of (standard model) light shining through a wall. The crosses denote the static magnetic field. Left: Individual higher order process producing light particles from a photon. Right: How light can traverse a wall (the black bar represents the wall, the hatched bubbles represent the complex higher-order physics described on the left side).

charge), which the second one can couple to. However, this constraint can be circumvented by the generation of virtual particles, which may mediate an interaction that cannot take place in lowest order. A typical example was already given in the previous section, where a virtual electron/positron pair mediated an interaction between two photons, which do not interact in lowest order. Another example is shown in Fig. 3.4. Here a virtual electron-positron pair couples to another virtual Z^0 boson. The Z^0 subsequently can decay under neutrino emission, which in sum leads to the decay of a photon inside a magnetic field into a pair of neutrino and antineutrino [119]. Energy conservation should be fulfilled already for photon energies of 1 eV (i.e. NIR spectral range) because the neutrino is supposed to be an extremely light particle (see [9]) and the other particles involved are only virtual ones. Of course, the same process works in the backward direction.

Neutrinos couple only to the weak force and therefore traverse ordinary matter nearly unharmed. One might think of placing a wall in the middle of the magnetic field region, which blocks the light completely but does not significantly hamper the (anti-)neutrino. A fraction of them will convert back into photons in the magnetic

3. ALPS I project - Particle physics with high-power green light

field on the other side of the wall and the experimenter will observe LSW. This light will show the same characteristics (polarization, wave vector orientation and length) as the light in front of the wall [4], merely strongly attenuated. It should be noted, that also this process can be understood completely within the SM without any extensions or new physics! Unfortunately, it involves a chain of virtual particles and therefore its predicted probability is so small that it is not reasonable to expect its experimental observation in the foreseeable future. [119].

But in principle the effect of LSW can be used as a probe for unknown particles. Let's imagine for a while that an experimenter lives in a world without knowledge about the weak force and thus without knowledge about the Z^0 boson and the neutrino. This experimenter might conduct decay experiments to learn about the existence of this force. He might also does particle accelerator (or collider) experiments. Moreover, this experimenter might conduct an experiment as is sketched in Fig. 3.5 and try to observe light behind the ideally non-transparent wall placed in the middle between the magnets. If he could measure some signal, he would have obtained a direct experimental evidence that his model of the world is incomplete because there must be additional particles to explain this effect. By performing precise measurements of the probability to measure a photon, of the dependence on the magnetic field strength and other parameters he might in the end come to the conclusion that he has to include a new force into his model of the world.

An interesting point lies in the fact that the Z^0 boson is very massive. Therefore a collider experiment with the aim to produce this particle would need to provide a large amount of energy in the collision. Thus low-energy physics with photons can in principle complement collider experiments in probing high-energy physics. Low conversion probabilities in an LSW experiment might be overcome by the extraordinary high flux of primary particles compared to a collider experiment.

3.1.5. Problems of the Standard Model

Despite its success the SM is expected to be incomplete for several reasons. On the one hand its structure appears arbitrary [58]. It consists, for example, of bosons and

3.1. Particle physics with laser light

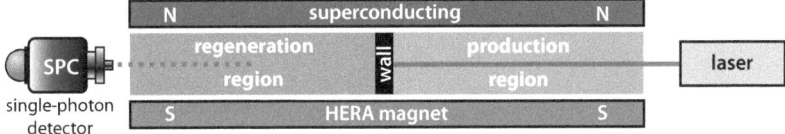

Fig. 3.5: Simple schematic overview of an experiment that uses the effect of LSW to search for hypothetical particles. Green lines and dots depict laser light. The red and green fields depict the poles of a magnet. Some hypothetical particles would need such a symmetry breaking field for their production from laser light and some would not.

fermions, some being very heavy and others completely massless, with a quantized charge property and exactly three different color charges. On the other hand there are observations, which cannot be explained at all in terms of the Standard Model. Some of them are listed in the following.

Dark matter and dark energy

Very much information about the composition of the universe can be obtained from precise measurements of the cosmic microwave background (CMB), which can be understood as the remaining light of the Big Bang, red-shifted to very low frequencies. It resembles a black body emission spectrum with small spatial anisotropies, which must be met by any model of the evolution of the universe [19].

The WMAP mission measured these spatial anisotropies with very high precision [18]. Since the publication of the seven year results of WMAP there is little doubt left that only 4.6 % of the universe is made up of ordinary matter [70]. Additionally there is 23 % so-called dark matter, which means matter of unknown kind not made up of atoms, that does not interact with electro-magnetic radiation (thus it can also be utterly transparent) and only very weakly with other matter via gravitation [59]. Its existence was first proposed by Zwicky in 1933, who observed a mismatch between the apparent velocities in the Coma cluster and its observable luminous media [160].

The only standard model particle that has similar features and can exist in free space for long times is the neutrino (see Fig. 3.1). But from current neutrino mass limits and WMAP data it can be concluded that only a small fraction of the 23 % may consist of neutrinos [20].

The remaining 73 % are made up of so-called dark energy. Dark energy is a form of energy, which tends to increase the expansion rate of the universe. It resembles the effect of Einstein's cosmological constant and its corresponding energy scale is on the order of a millielectron volt [109]. Here the situation becomes even worse because the standard model of particle physics has no means at all to explain this.

Due to the accumulated observational evidence for the existence of dark matter and dark energy especially from the WMAP data the current Cosmological Standard Model takes their existence as a given, hence it is named LCDM model for Lambda-Cold-Dark-Matter (Lambda is the abbreviation for Einstein's cosmological constant).

Effective number of neutrino species

If the universe for any reason contains a large reservoir of one or more species of relativistic light particles, which hardly interact with electromagnetic radiation or other matter, this reservoir would contribute to the energy density of the universe. In the SM such a reservoir might be filled by neutrinos and thus the corresponding energy density is generally parametrized by the effective number of neutrino species N_{eff}, which fill this reservoir [55].

This energy density would again impose some feature on the the CMB, which should be observable. In the SM a value of $N_{\text{eff}} = 3.0$ follows from the existence of three neutrino species in this model [76]. But combination of data from the seven year WMAP results, from latest distance measurements in the distribution of galaxies [112], and from Hubble constant measurements [121] resulted in a value of $N_{\text{eff}} = 4.34 \pm 0.9$ [70]. Thus it seems possible that additional neutrino-like particles beyond the SM exist in the universe. The Planck satellite will significantly reduce the uncertainty in this value.

Strong CP problem

It is generally known that the weak interaction violates the CP symmetry, i.e. the symmetry of a process under inversion of charge and inversion of all three spatial coordinates [21]. This is especially interesting because there exists no experimental evidence for CP violation of the strong interaction although the underlying well-tested theory of quantum chromodynamics (QCD) clearly allows for that [27]. This is not impossible. But it appears to be very unlikely from a theoretical point of view because it requires that two contributions from very different physical origins in the Standard Model, which are not necessarily small, sum up to give a very tiny value of $< 10^{-10}$ [108, 107]. While a small but non-vanishing electric dipole moment of the neutron would be a hint for CP violation of the strong force, very precise measurements have put these stringent constraints on that.

A possible solution to this problem (maybe even the most likely one), which was suggested by Peccei and Quinn, is to extend the Standard Model by an additional symmetry, which is spontaneously broken at low energies [107]. The breaking of symmetries can be understood in analogy to the thermodynamics of a medium, which crosses a second-order phase transition where its originally single state bifurcates into two distinct states, whose selection depend on an additional parameter (e.g. water cooling below its critical point) [113]. This new symmetry is called PQ symmetry after its inventors.

Gravity

As is obvious from the schematic of the SM in Fig. 3.1 it does not account for gravity. This is partly caused by the fact that there is no convenient and non-divergent way to quantize the field equations of Einstein's General Relativity. Due to the weakness of this force this is not important from the particle physics experimentalist's point of view. A quantum theory of gravity would simply have no known observational consequences [113]. These would not become relevant until the Planck mass scale is

3. ALPS I project - Particle physics with high-power green light

reached at

$$m_{\text{planck}} = \sqrt{\frac{\hbar c}{G}} \approx 1.2 \times 10^{19} \frac{\text{GeV}}{c^2} \quad .$$

At this scale it seems probable that any model based on continuous spacetime will break down. A suitable model for the description of spacetime at this scale might be string theory [113]. It substitutes the point-like particles of the standard model by extended strings. Since these strings cannot be exactly localized spacetime itself becomes nonlocal and distances much smaller than the Planck length $1/m_{\text{planck}}$ might not exist. Furthermore additional dimensions have to be added to the model to make it consistent with the observable world.

For comparison, the maximum mass scale that will be reached by the LHC once it will be at full performance is about $1 \times 10^4 \frac{\text{GeV}}{c^2}$.

3.2. Hypothetical hidden sector particles coupling to photons

3.2.1. Three kinds of hypothetical particles

At least some of the problematic aspects of the SM, which were listed at the end of the previous section might be solved by the postulation of extensions to the Standard Model. Most of these require the existence of new light particles (with masses below 1 eV), which often interact only gravitationally with Standard Model matter or through higher order processes involving virtual particles. Therefore such a new particle is denoted as a WISP. Especially string theory motivated Standard Model extensions are prone of postulating WISPs [68, 6, 107].

Such particles are often also considered as populating a so-called hidden sector. This means that their particle properties do not match those of the SM particles (e.g. electric charge). As was explained in the context of 3.1.4 interactions between such particles and Standard Model particles are therefore only possible via higher-order processes involving maybe heavy virtual mediator particles or via gravity. This is the reason for their very low interaction probabilities with ordinary SM matter.

3.2. Hypothetical hidden sector particles coupling to photons

In the following those WISP species are introduced shortly, which were of special interest in the scope of this thesis. For all of them the conversion probability of a laser photon (γ) into a WISP (ϕ) and back in a symmetric LSW experiment with a coherent production light beam is given by [44]

$$\mathcal{P}(\gamma \to \phi \to \gamma) = \frac{\delta^4 L^4}{16\omega^4} \operatorname{sinc}^4\left(\frac{M^2 L}{4\omega}\right) \quad , \tag{3.1}$$

$$N_r = \mathcal{P}(\gamma \to \phi \to \gamma) N_p \quad .$$

Here $M^2 \approx 2\omega^2 \Delta n + m^2$, ω is the photon energy, $\Delta n = n - 1$ denotes the deviation of the refractive index from its vacuum value, m is the WISP mass, L is the length of the production and regeneration region and δ depends on the WISP species. N_r denotes the number of regenerated photons in the regeneration region in a certain time interval and N_p the number of photons in the production region travelling into the direction of the detector within the same interval. The sinc-function is defined as $\operatorname{sinc}(x) = \sin(x)/x$. It should be noted that Eq. (3.1) and the definitions of δ given below are in natural units with all parameters either dimensionless or in units of electron volts except g_+ and g. which will be in units of $\frac{1}{\text{GeV}}$.

As all WISP species listed in the following have a non-vanishing rest mass, energy and momentum conservation require that the WISP field has a longer wavelength than the producing laser field. This means that a photon field and a WISP field will be running out of phase although they are travelling collinearly. Thus after a certain propagation distance destructive interference will occur in the conversion from one field to the other. This is the reason for the sinc() function in Eq. (3.1), which periodically reduces the conversion probability to zero over the propagation distance. This effect is exactly the same, which occurs in the plane-wave model of imperfect phase-matched SHG (see chapter 2 and [26]).

Axion-like particles

An axion-like particle (AP) is a spin 0 particle. Therefore they can be produced from spin 1 laser photons only inside a magnetic field where the laser photon can interact with a virtual photon of the magnetic field to conserve spin. Fig. 3.6 depicts the

117

3. ALPS I project - Particle physics with high-power green light

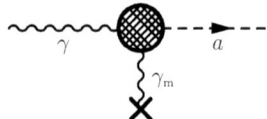

Fig. 3.6: Feynman diagram depicting the conversion of a laser photon (γ) into an AP(a). The hatched bubble represents unknown physics in analogy to Fig. 3.4. The cross denotes the static magnetic field.

corresponding Feynman diagram. Depending on the model either scalar particles are predicted (e.g. [148, 32]), which can be produced in a magnetic field oriented parallel to the laser electric field polarization with $\delta = g_- B \omega$, or pseudoscalar particles are produced (e.g. [146, 49]), which need a magnetic field oriented perpendicular to the electric field polarization and have $\delta = g_+ B \omega$. Furthermore here $L = L_m$ is set because for the conversion probability into or from this species only the propagation distance in the magnetic field is important.

The QCD axion

The QCD axion is a special example of a scalar axion-like particle. It arises from the postulation of the PQ symmetry, which was explained above to be a likely solution of the strong CP problem. Many models relate its mass to its coupling constant in a typical way leading to a band in parameter space with thickness of ca. one to two orders of magnitude in g_- and with its upper limit at [42, 102]

$$g_- \approx 10^{-12} \frac{m}{1\,\text{meV}} \quad .$$

Astrophysical observations imply that if the QCD axion exists, its mass should be below 10^{-2} eV [68, 130], which makes it a typical WISP. As the QCD axion is only a special axion-like particle it also needs a magnetic field to be produced. The Feynman diagram and expression for δ are identical to that one for APs.

3.2. Hypothetical hidden sector particles coupling to photons

Fig. 3.7: Feynman diagram depicting the conversion of a laser photon (γ) into a MHP($\gamma\prime$). The 'ball' represents unknown physics.

Massive hidden sector photons and mini-charged particles

The massive hidden sector photon (MHP) and the mini-charged particle (MCP) often arise from string theory motivated extensions of the standard model. MHPs are spin 1 particles and thus do not need a magnetic field to be produced from ordinary laser photons by a process called kinetic mixing. Similar to photons in the electro-magnetic sector of the SM MHPs represent the force carriers inside their additional hidden sector [119]. As they have rest mass they are dubbed here massive hidden sector photons [6, 7]. Other names used for them are hidden photons or paraphotons [6]. The corresponding Feynman diagram is shown in Fig. 3.7. The WISP species dependent parameter from Eq. (3.1) for MHPs is given by $\delta = \chi m^2$.

MCPs are the fermions of the hidden sector. They carry the hidden charge the MHPs couple to [119]. By the same kinetic mixing process they will also acquire a small fractional electric charge. Due to this electric charge they will be pair produced if laser photons travel in a magnetic field. They will act as virtual mediators coupling laser photons to hidden sector photons, which can either be MHPs or even hidden sector photons without rest mass [1, 54, 7]. In this process the magnetic field is not necessary for spin conservation but simply enlarges the conversion probability [119]. This process is depicted in Fig. 3.8. The WISP species dependent parameter δ is complex in this case and omitted here. It is given for instance in [28]. Its scale is further on denoted as Q.

3. ALPS I project - Particle physics with high-power green light

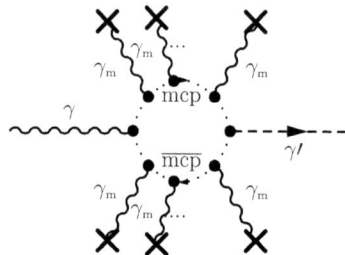

Fig. 3.8: Feynman diagram depicting the conversion of a laser photon (γ) into a hidden sector photon ($\gamma\prime$) via a virtual MCP loop. The crosses denote the static magnetic field.

3.2.2. WISPs as answers to puzzling astrophysical observations

Several astrophysical observations exist, which are difficult to understand within Standard Model physics. For some of them extensions of the Standard Model based on the existence of certain WISP species could be developed, which yielded a more precise theoretical description of the measured data. Examples are:

- If any QCD axion would be found this would proof that the strong CP problem is in fact due to the existence of a PQ symmetry as explained above. A QCD axion with a mass in the region of $1\,\mu$eV would be a very good candidate for cold dark matter in the universe [130, 131]. The ADMX experiment searches for specifically this type of QCD axion [10]. Moreover, cooling models of white dwarfs fit observational data significantly better if the emission of QCD axions with masses of $2-6$ meV and thus a coupling constant on the order of $10^{-13}\,\frac{1}{\text{GeV}}$ is assumed [66].

- The MAGIC telescope measured a surprisingly high transparency of the universe for very high energy radiation around 500 GeV hardly consistent with usual models of the extra-galactic background light [8]. This measurement can be explained much better if the oscillation of high energy photons into very

120

light APs and vice versa in the intergalactic magnetic field is assumed [122]. Such an AP would have a mass smaller than $m = 10^{-10}$ eV and a coupling constant in the range $10^{-13} \frac{1}{\text{GeV}} \leq g \leq 10^{-11} \frac{1}{\text{GeV}}$. Moreover, the x-ray emission spectrum of the sun and the process that strongly heats its corona is not understood. If APs exist they would contribute considerably to the abrupt temperature gradient between the solar corona and the sun's interior [159].

- The increase of the effective number of neutrino species $N_{\nu,\text{eff}}$ found by an analysis of the seven year WMAP data combined with other recent results (see [70] and explanation above) can be explained in such that MHPs with coupling constant $\chi \approx 10^{-6}$ formed a so-called hidden microwave background, which contributes to the value of $N_{\nu,\text{eff}}$ [67].

3.3. State of the art of LSW experiments in the world

The results of the first large-scale LSW experiment ever were published in 1993 by the BFRT collaboration at the Brookhaven National Lab New York City, USA [29]. Interestingly, the BFRT experiment already utilized a delay line to enlarge the propagation length of the laser pulses inside the magnetic field region.

At the launch-time of the first phase of the 'Any Light Particle Search' (ALPS) experiment at DESY, Hamburg there were four other large-scale LSW experiments active in the world. These were

- the 'Gamma to milli-eV' (GammeV) experiment at Fermilab Batavia, Illinois, USA,

- the 'Optical Search for QED vacuum magnetic birefringence, Axions and photon Regeneration' (OSQAR) experiment at CERN Geneva, Switzerland,

3. ALPS I project - Particle physics with high-power green light

- the 'Birefringence Magnetic du Vide' (BMV) experiment at LNCMI Toulouse, France, and

- the 'Light Pseudoscalar and Scalar Search' (LIPSS) experiment at Jefferson Lab Newport, Virginia, USA.

exp.	laser technology	total prod. photons	magnet $B \cdot L_m$	$g_. \times 10^7\,\text{GeV}$, (95 %)
GammeV	20 Hz, 532 nm, $P_{avg} = 3.2\,\text{W}$	6×10^{23}	$2\times 15\,\text{Tm}$	2.9
OSQAR	CW (single-pass), 514 nm, $P_{avg} = 18\,\text{W}$	not known	$2\times 136\,\text{Tm}$	3.4
BMV	0.28 mHz, 1064 nm, $P_{avg} = 0.36\,\text{W}$	4×10^{23}	$2\times 4.4\,\text{Tm}$	11.0
LIPSS	75 MHz, 935 nm, $P_{avg} = 180\,\text{W}$	5×10^{25}	$2\times 1.8\,\text{Tm}$	14.0

Table 3.1: List of all large-scale LSW experiments, which were active when work at ALPS I (phase 1) started. The applied technology and best sensitivity is shown for GammeV [33], OSQAR [13], BMV [46] and LIPSS [5] respectively. The OSQAR collaboration did not publish sufficiently detailed results to deduce their number of photons in the production region.

Tbl. 3.1 gives an overview over their applied technology and highest achieved sensitivity in the scalar AP coupling constant $g_.$. Obviously the GammeV experiment was the most sensitive one. Most experiments were based on pulsed lasers. The BMV experiment was mainly set up as a probe for the effect of vacuum birefringence, which was explained in the context of Fig. 3.3. This effect was also searched for by the PVLAS experiment [156]. From the results of such experiments also upper limits for the coupling constants of APs, MHPs and MCPs can be derived if the induced vacuum refractive index is interpreted as oscillation of photons into WISPs and vice versa.

Apart from these LSW experiments upper limits for the coupling constants also come from cosmological and stellar evolution requirements. Additionally, experiments, which try to detect WISPs of astrophysical origin like a future SHIPS in Hamburg and CAST at CERN determine upper limits. On the one hand those upper limits are usually more stringent because they do not have to produce the WISPs first. On the other hand they rely on a necessarily imperfect knowledge about the exact astrophysical processes involved.

Moreover, upper limits can also be derived from the non-observation of violations of fundamental principles like the Coulomb law.

A complete compilation of upper limits from all types of experiments just given for all WISP species of interest in the scope of this thesis can be found in [68] and references therein.

3.4. The first LSW experiment with production resonator

To continuously increase the sensitivity of an LSW experiment one has to improve all vital subsystems. In the GammeV experiment these subsystems were the pulsed primary laser, the magnets and the photomultiplier used as single-photon detector [33]. Apart from the cryogenic magnet technology, improvement of these subsystems would have been possible to some extend. But at some stage the chosen technologies themselves will impose limits on the achievable sensitivity.

The very high peak intensities and correspondingly large thermal gradients and induced nonlinear effects of pulsed lasers with high average power will tend to degrade their temporal pulse shape and lateral beam shape [74]. This would become

3. ALPS I project - Particle physics with high-power green light

problematic from two points of view. On the one hand typical accelerator magnets, which are necessary to search for certain WISP types, have rather small radii of their free aperture on the order of 20 mm. Due to the increasing diffraction of a Gaussian beam with decreasing power fraction η_{00} contained inside the TEM$_{00}$ mode (see section 2.1.2), the distortion of the lateral beam shape will limit the achievable magnetic field length L_m. On the other hand a benefit of pulsed lasers is the possibility to gate the detector, which strongly reduces their effective electronic dark noise. But if the temporal pulse shape gets worse the detector gate has to be kept open for longer periods of time.

Another limit will arise from the detector, if its dark noise depends on the size of its active area. This is for instance the case if the dark noise is dominated by thermal radiation from the surrounding field of view [143]. The same arguments just given imply that with growing η_{00} the impinging light beam can be focussed to smaller spot sizes on the detector, leading in such scenarios to lower dark noise.

In contrast to all other experiments specialized to the LSW effect the ALPS experiment decided to discard the possibility to gate the detector and instead to implement a CW laser source in combination with an optical resonator comprising the production region. In a first step the compatibility of this technology with the challenges of a large-scale LSW experiment was to be shown. This experiment is subsequently referred to as ALPS I (phase 1).

The implementation of a production resonator in an LSW experiment in principle allows to reduce the intensities and thermal gradients inside of the laser oscillator, which would be the main cause for a degraded output beam profile. The lower primary laser power is then resonantly enhanced by the cavity comprising the production region (also denoted as production cavity). Because this cavity is empty, one faces much less problems related to the high circulating power compared to the case where this power would be circulating inside a (necessarily not empty) laser resonator. Hence the eigenmode of the optical resonator will be close to an ideal TEM$_{00}$ mode, which relaxes the constraints from the small free aperture of the cryogenic magnets. Substitution of the gated detector by an integrating single-photon counter based on only a few pixels of a cooled CCD chip kept the dark noise low. The mode

3.4. The first LSW experiment with production resonator

filtering properties of the production cavity supported this tight focussing.

3.4.1. Design and experimental setup of ALPS I (phase 1)

The overall setup

Fig. 3.9 shows a rudimentary schematic overview of the experimental setup of the overall ALPS I (phase 1) experiment. The experiment was located in building 55 on the grounds of Deutsches Elektronen Synchrotron Hamburg, Germany (DESY). This is an industrial-type building, where the necessary mounting and cooling facilities for the heavy magnet were available. Laser, corresponding optics and control electronics were located inside of a big container (also called 'laser hut'), which was equipped with air conditioning and sources of filtered air, which provided protection against the building's ambient conditions. At the magnet's exit a cabinet was located, which contained the detection stage. A photo of the complete experiment is shown in Fig. 3.10.

The light from the primary infrared laser with a wavelength of 1064 nm was first converted to light with a wavelength of 532 nm in a single-pass SHG to match the sensitivity maximum of the employed single-photon detector. The converted light was directed to the production resonator. A weak so-called reference beam was split off and sent through a sealed tube along the outer side of the magnet to the detection stage.

A so-called production vacuum tube of length 6.3 m could be inserted into the laser-side half of the magnet to optimize WISP production efficiency (see Eq. (3.1)). The far end production resonator mirror was attached to its end. Approximately half of the production resonator length was permeated by a constant magnetic field of strength $B = 5.3$ T. This magnetic field was generated by a spare dipole of the former Hadron-Electron Ring Accelerator (HERA) particle accelerator at DESY. This 9.8 m long device generated its magnetic field at a full length of 8.8 m when supplied with a current of 6000 A and cooled to 4 K by circulating liquid helium. As it was designed to fit into an accelerator ring, it was bended with a radius of curvature of ≈ 588 m. The inner beam tube was made from titanium and had an

3. ALPS I project - Particle physics with high-power green light

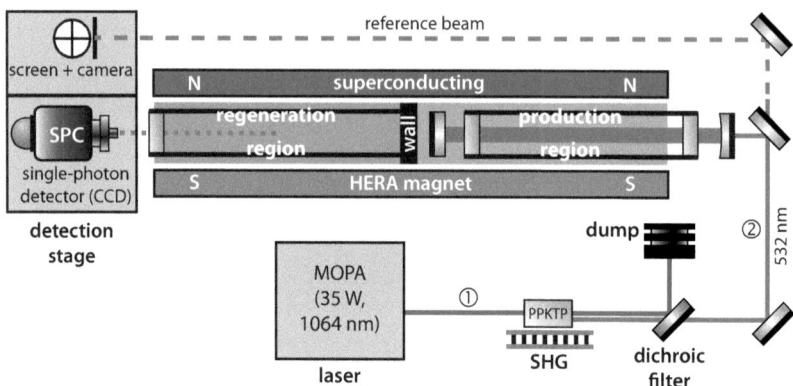

Fig. 3.9: Rudimentary schematic overview of the experimental setup of the whole ALPS I (phase 1) experiment in its initial phase. It is the only experiment specialized to detect LSW that utilizes a production resonator. Red lines (1) depict a laser wavelength of 1064 nm and green lines (2) of 532 nm.

overall length of 13.5 m because it had to extend beyond the connection chambers for the helium supply. It was kept at room temperature by a shielding technique called super-isolation and a steady flow of heated filtered nitrogen through its interior. Because the nitrogen could be injected and removed only at the ends of the magnet a temperature gradient inside the magnet could not be avoided. Therefore the magnet center was at $\approx$ 14 °C while the nitrogen injection end was significantly above room-temperature. During the time the vacuum tubes were inserted the center temperature even dropped below $\approx$ 8 °C as the insertion procedure hindered the nitrogen flow.

The second half of the magnet was used as regeneration region. For this a second removable vacuum tube of length 7.6 m could be inserted. It had a welded steel cap directed to the magnet center (representing the 'wall' of the LSW experiment) and a window directed to the detector stage. The various lengths, which were important

3.4. The first LSW experiment with production resonator

Fig. 3.10: Photo of the complete ALPS I experiment with laser hut (left, light gray), magnet (center, light yellow) and detector cabinet (right, dark green) as it was located in building 55 on the DESY grounds.

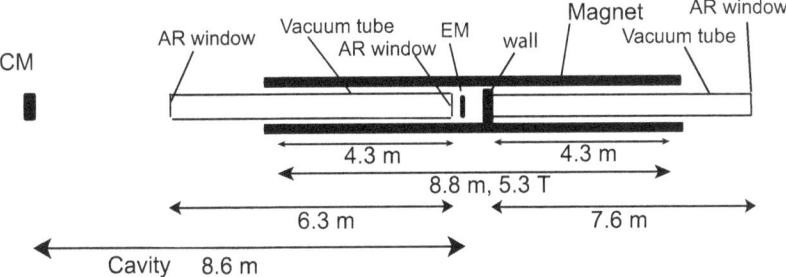

Fig. 3.11: Summary of important lengths of the ALPS I (phase 1) experiment. Here CM and EM denote the mirrors of the production resonator, AR means coated for low reflection and the 'wall' is denoted as absorber.

for the experiment are summarized in Fig. 3.11.

The detector stage consisted of a small aperture in front of a webcam acting as spatial reference for the weak reference beam and of a light-tight box containing the single-photon detector and some alignment and focussing optics. This box could be connected in a light-tight way to the regeneration vacuum tube such that no photons except those from black body radiation or regenerated light could hit the

3. ALPS I project - Particle physics with high-power green light

single-photon detector. Thermal radiation was calculated and tested experimentally to have no significant influence. To achieve low dark noise the single-photon detector was configured to collect data in time segments of 1 h each. Such a segment is denoted in the following as data frame.

The reference beam was also injected into the light-tight box through a very effective absorber such that only a few laser photons and no ambient light arrived at the detector. The full setup was arranged such that the regenerated photons would have been expected in a tiny so-called 'signal area' of only 45 µm × 45 µm on one half of the detector's active area while the residual photons of the reference beam could be detected in another small area on the other half. A more detailed description of the detection stage, the data taking and evaluation can be found in [44].

Injection stage and production resonator

In the scope of this thesis the optical injection stage, the production resonator and the corresponding control schemes of the ALPS I (phase 1) experiment were designed, set up, characterized and maintained. These subsystems are depicted in greater detail in Fig. 3.12 and are described in the following.

Laser. The beam tube inside the HERA dipole magnet was bent horizontally and therefore left a free horizontal aperture of only 14 mm in diameter when the vacuum tubes were in place. Moreover, the laser light should be resonantly enhanced within the production region. Especially the second point strongly constrained the beam quality of a suitable laser as the resonator is usually designed such that the transverse mode resonances do not degenerate. Furthermore, to efficiently increase the optical power with a cavity, one needed a continuous-wave laser that emits a single longitudinal and transversal mode with a much smaller linewidth than the one of the resonator. Thus, the laser source used for the ALPS I experiment was a narrow-linewidth master-oscillator power amplifier system (MOPA) operating at 1064 nm, which was provided by the Laser Zentrum Hannover (LZH). It is based on the system described in [48] developed for gravitational wave detectors like LIGO and GEO 600.

3.4. The first LSW experiment with production resonator

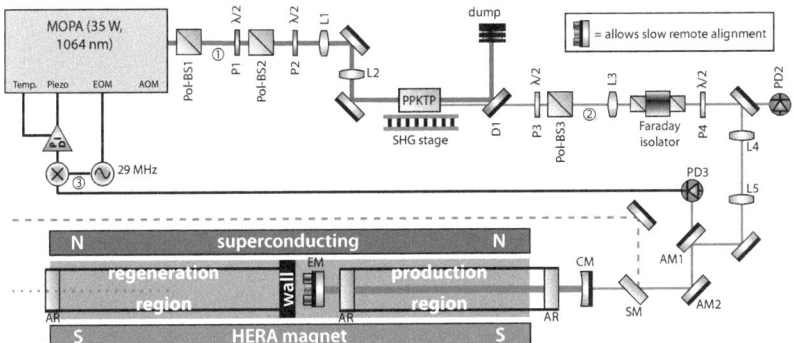

Fig. 3.12: Detailed schematic view of the optical setup of the injection stage and the production resonator of ALPS I (phase 1)including the frequency stabilization feedback loop. Red lines (1) denote a laser wavelength of 1064nm and green lines (2) of 532 nm. Black lines (3) denote electric signals. See text for description of components.

Stable narrow-linewidth emission was provided by a NPRO [72], with a spectral width on the order of 100 Hz within a measurement interval of 25 ms [35], and a long term frequency stability of 1 $\frac{\text{MHz}}{\text{min}}$, which acted as the master of the system and thus dominated its frequency characteristics. It was amplified to an available output power of $\approx 34\,\text{W}$, which were contained in a nearly diffraction limited beam with a fundamental transverse mode content of 95 % [44]. The MOPA was equipped with several frequency control elements. A piezo-electric transducer installed on the NPRO laser crystal allowed for a frequency shift of $\pm 120\,\text{MHz}$ with a response bandwidth from dc to $\approx 100\,\text{kHz}$. Slow frequency drifts could be compensated by controlling the crystal temperature with a tuning coefficient of $-3\,\frac{\text{GHz}}{\text{K}}$. Before amplification the NPRO beam was passed through an electro-optical modulator (EOM) and an acousto-optical modulator (AOM) giving the possibility to imprint an amplitude modulation (AM) or phase modulation (PM) on the infrared laser

3. ALPS I project - Particle physics with high-power green light

light.

The amplifier output beam had a polarization extinction ratio of more than 20 dB. A polarizing beam splitter (Pol-BS1) was used to further clean the emitted polarization state.

Single-pass SHG. In case of the effect of LSW the regenerated light behind the wall has the same characteristics as the laser beam in the WISP production region [4]. The quantum efficiency of the single-photon detector of the ALPS I experiment was strongly peaked in the green visible spectral region [114]. Therefore the infrared laser light was first converted from 1064 nm to green 532 nm light utilizing the nonlinear effect of SHG [25, 134].

The ALPS I (phase 1) experiment was meant as a test for the compatibility of a large-scale LSW experiment and long optical resonator. This test did not need a high incident power on the production resonator. Thus to avoid unnecessary complexity a single-pass SHG stage was implemented. As nonlinear material PPKTP was used, which assured sufficient single-pass conversion efficiency due to its high intrinsic nonlinearity. The crystal's dimensions were 1 mm × 2 mm × 2 cm and its nonlinearity was measured to be $d_{\text{eff}} = 7.9 \times 10^{-12} \frac{\text{m}}{\text{V}}$. Identical to the PPKTP crystal of the high power SHG described in 2.3 the crystal was mounted inside a self-made oven to stabilize its temperature at around 38 °C in order to maintain phase matching of the fundamental and harmonic waves. Two lenses (L1, L2) were used to focus the infrared beam into the crystal to a waist size of 135 µm. This waist size was chosen as a compromise between highest theoretical conversion efficiency (obtained for Boyd-Kleinman focussing to a waist of 26 µm), the risk of damaging the crystal and degradation of the harmonic beam shape (both originating from too high intensities inside the crystal).

The input polarization was adjusted to maximum conversion efficiency by $\lambda/2$-plate (P2). The input power level of the infrared beam was set via a variable attenuator consisting of another $\lambda/2$-plate (P1) and a polarizing beam splitter (Pol-BS2). Behind the oven the converted green light was separated from the infrared by means of a dichroic mirror (D1), which was followed by a variable attenuator for the

3.4. The first LSW experiment with production resonator

green light beam (P3, Pol-BS3) and a collimating lens (L3). Then the light passed a Faraday isolator to protect the nonlinear crystal from back-reflections from the following optical setup, especially in the case of a non-resonant production cavity. The amount of power incident on the cavity was monitored with photodetector PD2. The polarization state of the laser light entering the cavity could be adjusted by a $\lambda/2$ plate (P4). By this a linear polarization state with an arbitrarily adjustable polarization angle relative to the magnetic field direction was realized, which was important to be able to search for all types of WISPs, which were of interest.

Production resonator. As production cavity of the ALPS I experiment a linear resonator consisting of two mirrors was realized. Its design was constrained by several aspects. First, WISPs are produced most effectively in vacuum, and hence as much as possible of the resonator had to be evacuated. Second, the site at DESY did not allow for two magnets in a row. Thus the production and regeneration regions of the LSW experiment had to be located inside the same magnet. This required that the far end mirror of the cavity had to be placed at the center of the HERA magnet within a tube with only 33 mm inner diameter. Third, the bending of the magnet bore resulted in a cat's pupil-shaped free aperture of only 28 mm height and 14 mm width.

To keep the complexity of the optical setup as low as possible, both resonator mirrors were located outside of the vacuum tube. The input coupler (CM) was mounted inside a commercially available manually adjustable mirror mount, which was placed on the optical table of the injection stage in front of the magnet. The far end mirror (EM) was mounted on a self-made non-magnetic remotely adjustable mirror mount. It was attached to the head of the production vacuum tube and both were inserted into the magnet. The tube had windows on each end, which were coated to reduce their reflectivity (AR). The pressure inside this tube was routinely kept below 10^{-5} mbar, which kept the influence of the refractive index of the rest gas far below significance. The design and construction of vacuum tube and far end mirror mount were not part of this thesis.

The distance of the resonator mirrors resulting from this design was 8.62 m of

3. ALPS I project - Particle physics with high-power green light

which 6.3 m were occupied by the vacuum tube of which 4.3 m were exposed to the magnetic field. The mirror substrates were polished to a planarity of $\lambda/4$ according to the international standard MIL-O-1380A and afterwards were coated by the electron beam deposition coating technique [53], which could be obtained fast and at reasonable costs. Their approximate quality was known from other experiments with comparable mirrors at a wavelength of 808 nm, which resulted in losses per mirror by absorption and scattering of 0.14 %.

A straight-forward way to increase the number of regenerated photons due to the LSW effect is to increase the number of photons in the production region. This is done by optimizing the power buildup of the production resonator. As was explained in 2.1.3 in case of an empty cavity this optimization is done by first minimizing the fractional passive losses A_p and then choosing the transmission of the input coupler to match the remaining value of A_p (i.e. realizing an impedance matched cavity). Therefore the far end mirror was chosen to be highly reflective with a measured residual transmission of $T_{out} = 170$ ppm. The production resonator comprised the two windows of the vacuum tube. Thus the circulating light had to traverse eight surfaces coated to reduce reflection. Their remaining fractional reflection was stated by the manufacturer to be ≤ 0.3 %. From these values a design value for the transmission of the input coupler leading to an impedance matched cavity could be estimated to be $T_{in} \leq 8 \cdot 0.3\% + 2 \cdot 0.14\% \approx 2.7\%$. Accurate measurements of the transmission of the purchased input coupler resulted in a power transmission of $T_{in} = 2.3\% \pm 0.14\%$. From the values stated here a design value for the linewidth could be calculated as $FWHM = 130$ kHz. Due to the large length of the resonator its free spectral range could be determined with high precision by an ordinary distance measurement of the mirrors to be $FSR = 17.4$ MHz.

For the given mirror distance several stable resonator designs were possible that differ in the radii of curvature of the cavity mirrors. They were simulated by application of the ABCD matrix formalism [129, 75]. For the production resonator of ALPS I a plano-concave design was chosen with a plane far end mirror and a concave input coupler with $ROC = -15$ m. As is shown in Fig. 3.13 the resulting fundamental cavity eigenmode needs a free circular aperture of less than 6 mm di-

3.4. The first LSW experiment with production resonator

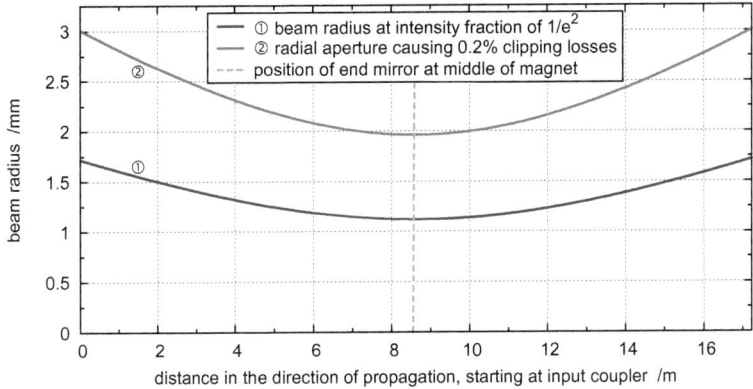

Fig. 3.13: Evolution of the beam radius of the TEM$_{00}$ eigenmode of the optical resonator together with the aperture radius, at which 0.2 % of the mode's power would be clipped. The position of the plane far end mirror is shown as a vertical line. Clearly, the beam size is always well below the minimum aperture of our production vacuum tube of 14 mm horizontal diameter.

ameter to keep power losses per round trip due to clipping below 0.2 %. Thus the free aperture of the magnet of 14 mm in horizontal direction was not limiting the power buildup.

Determination of the circulating power during WISP search. If a search for the LSW effect was conducted, the regeneration tube with the 'wall' attached to its end was inserted into the magnet, which blocked the transmitted beam of the production resonator. Right behind the regeneration tube at the magnet's exit the detection stage was placed and it took data for 1 h. Only after that detection stage and regeneration tube could be removed to have access to the transmitted light from EM again. It appeared certain that the production resonator would not have been stabilized to its maximum power buildup for this whole period of time.

133

3. ALPS I project - Particle physics with high-power green light

Correspondingly, while searching for WISPs, the circulating power inside the cavity had to be monitored by other means than via the transmitted power to be able to determine the effective sensitivity of the overall LSW experiment afterwards. This was accomplished by monitoring the incident power with PD2 and the power reflected from the cavity with PD3. Eq. (2.16) gives an expression for the circulating power P_{circ}. Explicitly allowing for static and dynamic mismatches of frequency, alignment or focussing between incident mode and cavity eigenmode the expression can be rewritten as

$$\frac{P_{\text{trans}}}{T_{\text{out}}} = P_{\text{circ}} = PB_{\text{max}}\,\eta_{00}^{(\text{opt})}\,P_{\text{inc}}\,f_{PB}(\Delta\nu)\,f_\eta(\Delta\eta_{00}) \qquad (3.2)$$
$$= P_{\text{circ}}^{(\text{opt})}\,f_{PB}(\Delta\nu)\,f_\eta(\Delta\eta_{00})$$

where Eq. (2.19) and the definitions

$$f_{PB}(\Delta\nu) = \frac{1}{1 + \left(\frac{\Delta\nu}{FWHM/2}\right)^2}\;,$$

$$f_\eta(\Delta\eta_{00}) = 1 - \frac{\Delta\eta_{00}}{\eta_{00}^{(\text{opt})}}$$

were used. Here $\eta_{00}^{(\text{opt})}$ denotes the value of η_{00} for a given 'optimal' alignment and $\Delta\eta_{00}$ the (possibly negative) deviation from this starting value.

The transmission factor of the production resonator end mirror was precisely known and thus prior to insertion of the regeneration vacuum tube the value of $PB_{\text{max}}\,\eta_{00}^{(\text{opt})}$ could be determined easily by measuring the incident and transmitted power in a well-aligned situation ($\Delta\eta_{00} = 0$) with stabilized power buildup ($\Delta\nu \approx 0$). The WISP search was performed in time segments of 1 h, during which the value of $PB_{\text{max}}\,\eta_{00}^{(\text{opt})}$ could be assumed to be constant. Based on this assumption the value of P_{circ} with inserted regeneration vacuum tube could be obtained from measuring the incident power and the relative change of the reflected power. Eq. (2.16) with the definitions around Eq. (3.2) gives

$$\frac{P_{\text{refl}}(\Delta\nu, \Delta\eta_{00}) - P_{\text{refl}}(\Delta\nu = 0, \Delta\eta_{00} = 0)}{P_{\text{refl}}(\Delta\nu = \infty) - P_{\text{refl}}(\Delta\nu = 0, \Delta\eta_{00} = 0)} = 1 - f_{PB}(\Delta\nu)\,f_\eta(\Delta\eta_{00}) \quad . \qquad (3.3)$$

The continuous measurement of the reflected power also allowed for repeated compensation of slow alignment drifts of the incident mode to the eigenmode, while

3.4. The first LSW experiment with production resonator

searching for WISPs. The focussing did not show any significant drifts on this timescale.

Where would regenerated light have hit the detector? As long as there was no regeneration vacuum tube inserted into the magnet, the residual transmission through the far end mirror of the production resonator could in principle be used to simulate a beam of regenerated photons for the characterization of the detection stage. It also allowed to determine the exact position of the tiny signal area where during a real measurement a hint to regenerated light was searched for. However, this assumption was only valid as long as the transmitted beam did not undergo any significant refraction by wedged optics, which would have caused him to travel along another path than WISPs, which do not interact with matter. From Snell's law and usual trigonometrics the misalignment of the beam behind an angled and/or wedged optic was calculated. For a plane parallel optic, which is placed under a small angle of incidence α only a beam displacement Δx occurs given by

$$\Delta x = d \frac{\sin\left(\left(1 - \frac{n_0}{n_m}\right)\alpha\right)}{\cos\left(\frac{n_0}{n_m}\alpha\right)} \approx d\alpha \left(1 - \frac{n_0}{n_m}\right) \quad . \tag{3.4}$$

This is different in case of an optic with a small wedge angle β, which is placed under a small angle of incidence of α. In this case the displacement Δx just calculated occurs again and also an angular misalignment $\Delta\Theta$ arises, which is given by

$$\Delta\Theta = \frac{n_m}{n_0}\left(\beta + \arcsin\left(\frac{n_0}{n_m}\sin(\alpha)\right)\right) - \beta - \alpha \approx \left(\frac{n_m}{n_0} - 1\right)\beta \quad . \tag{3.5}$$

In both equations n_m is the refractive index of the optic's material, n_0 the refractive index of the surrounding medium and d denotes the thickness of the optic. While the displacement was calculated to be of no importance in the ALPS I (phase 1) experiment, the angular deviation would have been amplified by the long propagation distance through the magnet. Right in front of the single-photon detector a focussing lens of diameter 1 inch was placed to concentrate the incident light onto the tiny signal area. The limit for the allowed wedge angle was derived from the requirement, that this lens had to be hit by the regenerated light within 0.5 cm from

3. ALPS I project - Particle physics with high-power green light

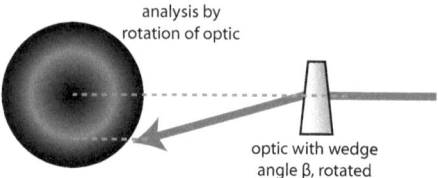

Fig. 3.14: The method applied to determine the wedge angle β of all optics, which were placed between production resonator eigenmode and single-photon detector.

its center. From this value and the experimental parameters of the overall experiment a maximum allowed wedge angle of the optics of 1 arcmin was calculated. Each optic, which was placed in the beam path from CM to the single-photon detector, was selected respecting this requirement. The test procedure applied was to mount the optic into a rotational stage, rotate it by 360° and watch the spot of the transmitted beam in the far distance (see Fig. 3.14). If the optic is wedged, the spot will move on a circle, whose radius and distance can be linked to its wedge via the equations given above. The optics, which were actually used in the setup, had wedges comparable to or smaller than 10 arcsec.

Section 2.1.3 explains that the power buildup is optimized when all of the incident power is contained inside the eigenmode of the resonator. In order to get as close as possible to this situation, an optimization of the spatial overlap of laser and resonator mode had to be performed. Once the cavity eigenmode was established with the help of the two adjustable cavity mirrors, the overlap optimization was done with the two beam shaping lenses L4 and L5 and the two alignment mirrors AM1 and AM2.

A pickoff beam (denoted as reference beam) from turning mirror SM was guided in a sealed tube along the magnet side and was used as spatial reference at the detector stage. Because the alignment and focussing of the incident beam was done only with optics in front of SM, this pickoff beam represented a good spatial reference for the

3.4. The first LSW experiment with production resonator

orientation and position of the resonator eigenmode once alignment and focussing of the incident beam to this eigenmode were optimized. Due to the sealing of its guiding tube by windows air currents did not harm this spatial reference. This feature was very helpful for the LSW experiment because once the regeneration vacuum tube with the 'wall' was inserted into the magnet the reference beam was the only possibility to test the correct alignment of the cavity eigenmode and thus of a hypothetical beam of regenerated photons to the tiny signal area of the single-photon detector.

Control schemes. Two feed-back control loops were necessary in the ALPS I (phase 1) experiment. The first stabilized the temperature of the PPKTP crystal inside its oven such that the phase matching of fundamental and harmonic waves was optimized. This control loop was already described in 2.3 and is not repeated here.

The second control loop had to minimize the mismatch between the frequency of the incident laser light and the TEM_{00} resonance frequency of the production resonator. This was necessary because the RMS value of this mismatch was large in a not stabilized (also called free-running) situation, which caused the power buildup to be low on average (see 2.1.3), reducing the number of photons in the production region and thus the sensitivity of the LSW experiment. To compensate for this RMS mismatch one could have either actuated on the production cavity length or on the MOPA frequency. To achieve a high response bandwidth typically only piezo-electric transducers are an option, which have a limited elongation of $\approx 1\,\mu m$ for manageable voltages of a few hundred volts. For a wavelength of 532 nm and a production cavity length of 8.6 m this translates into a range of such an actuator of $\approx 70\,MHz$, which would decrease with increasing resonator length. In contrast, as was described above actuation of the MOPA frequency had a constant range of 240 MHz. Therefore the latter was selected as actuator of the control loop.

The error signal of the control loop was generated by utilization of the well-known Pound-Drever-Hall (PDH) technique [22]. The modulation frequency was 29 MHz and these sidebands were imposed on the infrared light by the EOM, which was

3. ALPS I project - Particle physics with high-power green light

part of the infrared laser source. The light reflected by the cavity was detected by photodetector PD3 and the photocurrent was demodulated afterwards. To increase the SNR of the demodulated error signal, PD3 was equipped with an output, which realized a resonant readout of the modulation frequency.

One might wonder if the resonant SHG stage had an undesired impact on the modulation sidebands. By means of Eq. (2.22) together with Eq. (2.3) it is easily shown what happens if a phase modulated NIR beam with modulation frequency Ω and index M is converted. Its complex amplitude is described by

$$U_{0,1064} = U_0 \left[1 + i\frac{M}{2}e^{i\Omega t} + i\frac{M}{2}e^{-i\Omega t} + \mathcal{O}(M^2)\right] \quad ,$$

which is converted by the SHG effect into a beam, which is described by

$$U_{0,532} \propto U_{0,1064}^2 = U_0^2 \left[1 + iMe^{i\Omega t} + iMe^{-i\Omega t} + \mathcal{O}(M^2)\right] \quad .$$

The conversion process can be understood in analogy to a square detector. The sidebands of first order of the converted light are the product of a beat of the sideband with the carrier of the fundamental light, which increases their size relative to the carrier. In the experiments described here M was always small compared to unity and thus the only effect was a bigger demodulated error signal if equal detected optical powers are compared.

The demodulated error signal was then filtered by a PID controller. Its output was for one thing amplified to a dynamic range of ± 150 V and fed to the piezo-electric transducer of the infrared laser. And for the other thing it was low pass-filtered and fed to the crystal temperature of the NPRO of the infrared laser source.

In order to get well defined error signals out of the PDH scheme one had to assure that the resonance frequencies of higher order transversal modes with low index sum did not come closer to the TEM_{00} resonance than approximately three cavity linewidths. The radii of curvature of the resonator mirrors determine these resonance frequencies of the higher order modes [75]. The production resonator was thus simulated in advance by application of the ABCD matrix formalism (see [129, 75]) to find a suitable combination of round trip length and radii of curvature, while both parameters were at the same time constraint by the available space inside the magnet

3.4. The first LSW experiment with production resonator

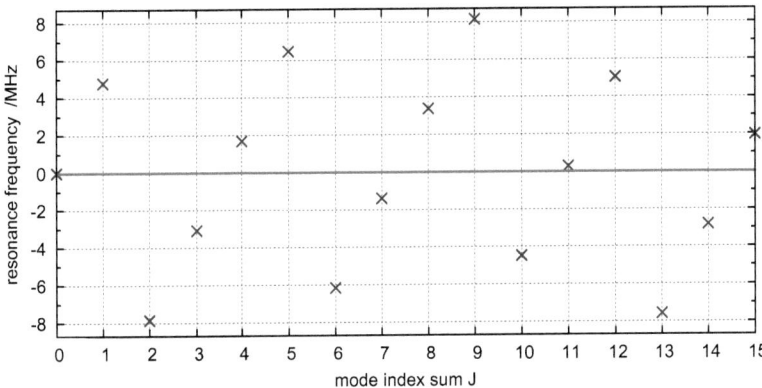

Fig. 3.15: Resonance frequency of transversal modes above its mode index sum as determined by the chosen mirror distance and radii of curvature of the ALPS I production resonator. Mode index sum J of a Hermite-Gauss mode TEM$_{nm}$ or a Laguerre-Gauss mode LG$_{pl}$ is calculated as $J = n+m = 2p + l$. The red line marks the TEM$_{00}$ resonance frequency as guide for the eye. The designed linewidth is considerably smaller than the size of the crosses. The full y-scale corresponds to one free spectral range.

and by such radii available for purchase. The chosen parameter set guaranteed that no relevant approach of resonance frequencies of higher order transversal modes to that one of the TEM$_{00}$ mode occurred including a mode index sum of 10. The overlap integral of the incident near Gaussian beam profile with higher order transversal modes decreased very rapidly with mode index sum. Correspondingly approaching modes with index sums above 10 were no problem. The simulation result is shown in Fig. 3.15.

To enable the achievement of a high duty cycle of the stabilized production resonator the control loop was equipped with automatic lock acquisition electronics. The loss of the stabilized state was detected as a jump of the power on detector

3. ALPS I project - Particle physics with high-power green light

PD3. The electronics reestablished the stabilized state again within $\approx 1\,\text{s}$.

3.4.2. Results of ALPS I (phase 1) and discussion

Single-pass SHG

The single-photon detector of the ALPS I experiment consisted of a cooled CCD chip, from which a rectangular area of $45\,\mu\text{m} \times 45\,\mu\text{m}$ was used for detection. Single-photon detection in the green visible spectral region is generally superior to NIR detection because silicon can be used as chip material, which shows a significantly higher quantum efficiency and lower dark current than chip materials for the NIR spectral region [41]. Due to long integration times of an hour the overall dark noise of the detector was dominated by this dark current. Thus, despite not aiming at highest sensitivities to WISPs with the current phase of ALPS I, the experiment was designed for a laser wavelength of 532 nm to avoid changes of fundamental design considerations in a following project phase.

At its first implementation the single-pass SHG scheme provided a stable harmonic output power of 1.3 W at a wavelength of 532 nm available behind D1 (transmission for harmonic wavelength $T \approx 94\,\%$) when the incident infrared power at the crystal entrance facet was set to its maximum power level of 31 W. This was acceptably close to its theoretical optimum, which can be calculated from Eq. (2.27) to be 1.6 W in this case. Its beam profile appeared to be gaussian. The theoretical harmonic power obtainable from insertion of a 5 cm long LBO crystal (as was applied in 2.4) into this setup would have amounted to only 0.55 W under best conditions including Boyd-Kleinman focussing.

Unfortunately, over the following hours the output power of the PPKTP single-pass SHG settled to a long-term stable value of 0.9 W of which 0.8 W were available at the production cavity input coupler. During this settling process the transversal harmonic beam shape became slightly non-Gaussian. The reason for these effects was expected to be due to thermal dephasing dynamics originating in time-dependent linear and nonlinear absorption processes as described in 2.2.4 and in 2.2.4. The slight deviation of the harmonic beam shape from a pure Gaussian lowered the cou-

3.4. The first LSW experiment with production resonator

pling efficiency to the production resonator's TEM_{00} mode but did not influence the overall experiment in any other way. Finally, the harmonic power incident on the production resonator was by far big enough to test the compatibility of a large scale LSW experiment with a long cavity.

Production resonator and frequency stabilization

Control loop bandwidth. From early inspection of the experimental site it was obvious that the amount of ambient acoustic and vibrational noise coupling to the injection optics and production cavity would become a critical aspect. To overcome these noise sources it was consequently important to achieve a large enough UGF of the control loop stabilizing the laser frequency to the production resonator. After a first rudimentary implementation of this loop its UGF could be measured and increased. Finally, a UGF of approximately 30 kHz was achieved and found to be sufficient. Fig. 3.16 shows an in-situ measurement of the open loop gain of the complete control loop at frequencies around its UGF. At lower frequencies its gain steadily increased up to far more than seven orders of magnitude. From this measurement it is obvious that the UGF could not be increased further due to a complicated structure of high frequency resonances of the piezo-electric transducer in the region from 100 kHz up to 300 kHz. It is hardly possible to compensate for this structure with analog electronics. A possible way to evade this limitation would have been the implementation of a third actuator to dominate the control loop at high frequencies, e.g. an EOM. But realization of this possibility would have meant significant additional efforts, which did not appear to be necessary for the desired stability, and were thus not pursued further.

Dynamic performance of control loop. After setting up a working frequency stabilization it could be utilized to gain more information about the frequency dependence and strength of the noise, which coupled to the length of the production cavity. Fig. 3.17 shows the linear spectral densities of the mismatch between laser frequency and cavity resonance frequency in an unstabilized (so-called free-running) and in a stabilized situation. Additionally the figure shows the estimated contribu-

3. ALPS I project - Particle physics with high-power green light

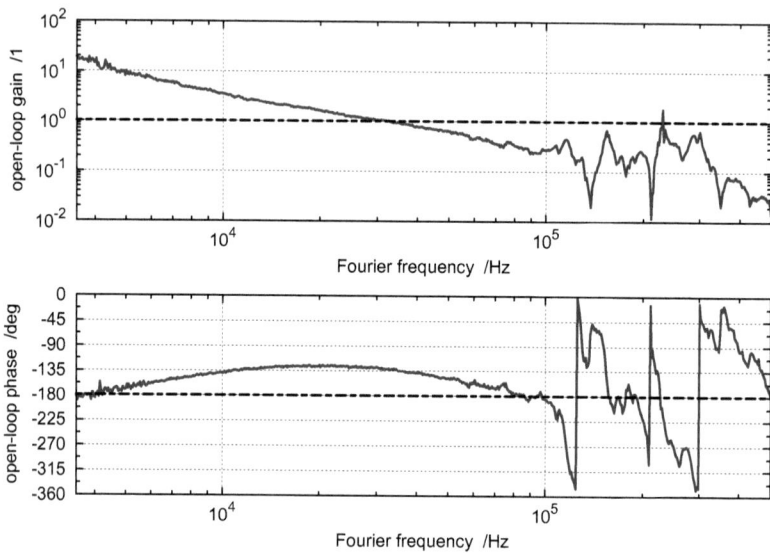

Fig. 3.16: In-situ measurement of the open-loop transfer function of the complete control loop stabilizing the laser frequency to one of the resonance frequencies of the production resonator. The gain popping above unity at $\approx 220\,\mathrm{kHz}$ is an artifact from the strong driving signal injected into the loop. The horizontal black broken line denotes a gain of unity (defining the UGF or control loop bandwidth) and a phase loss of 180° (theoretical limit of loop stability) respectively.

tion to this mismatch by the laser frequency noise alone, taken from [81]. In this publication a large number of devices were characterized, which are similar to the master laser of the infrared laser source used for the ALPS experiment. To judge their importance for the power buildup, the three curves just mentioned are normalized to the linewidth of the production resonator. As guide for the eye the figure also shows the UGF of the frequency control loop.

3.4. The first LSW experiment with production resonator

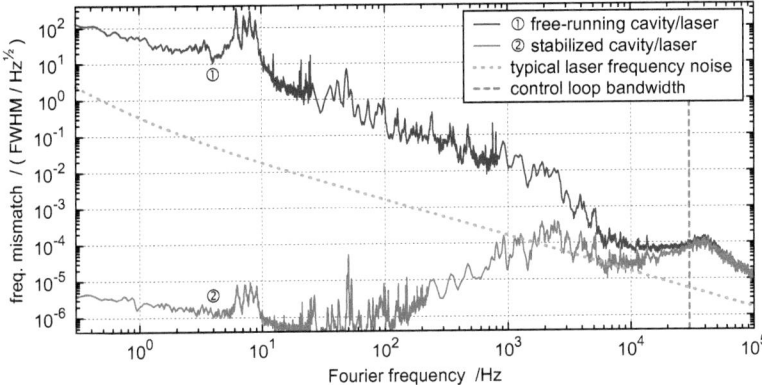

Fig. 3.17: Linear spectral density of the mismatch between laser frequency and production cavity resonance frequency in an unstabilized (so-called free-running) and a stabilized state. For comparison also the typical frequency noise of the laser alone is shown (taken from [81]). As guide for the eye the UGF of the frequency control loop is marked, too.

It is obvious that the overall mismatch was by far bigger than the contribution of the laser alone (note the logarithmic scale). Thus nearly all of the frequency mismatch was caused by noise of the production cavity resonance frequency, i.e. by its fluctuating microscopic length. A very prominent contribution comes from the structure just below 10 Hz. Correspondingly, if one wants to reduce the noise in the frequency mismatch, further stabilization of the laser frequency would not help much. Instead, the distance of the cavity mirrors would have to be isolated against the coupling of vibrational noise sources. The figure also shows that the frequency stabilization control loop worked well and suppressed the frequency mismatch starting from its UGF. The latter was deduced from the open-loop gain crossing unity in Fig. 3.16. The suppression of the frequency mismatch continued to acquire gain to lower Fourier frequencies reaching more than seven orders of magnitude of sup-

3. ALPS I project - Particle physics with high-power green light

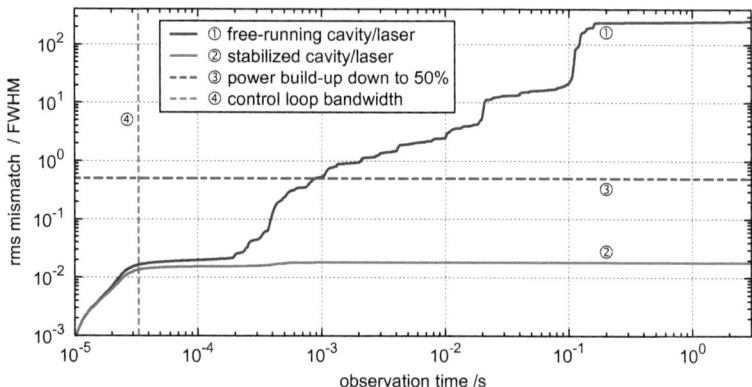

Fig. 3.18: The RMS mismatch between laser frequency and production cavity resonance frequency in an unstabilized (so-called free-running) and in a stabilized state is shown above the length of the observation time. As guides for the eye the UGF of the frequency control loop and the RMS mismatch corresponding to 50 % power buildup loss are marked, too.

pression at the leftmost shown Fourier frequency.

To judge if the free-running frequency mismatch is suppressed strong enough not to degrade the power buildup in the stabilized situation it is best to calculate the RMS frequency mismatch from the linear spectral density (LSD) given above. This was done in Fig. 3.18. Again everything is normalized to the production cavity *FWHM*. As guides for the eye also the control loop's UGF is shown and the value, which corresponded to a RMS reduction of the stabilized power buildup to half of its resonant value. Application of Eq. (2.19) shows, that at Fourier frequencies of 0.33 Hz corresponding to observation times of 3 s the stabilized power buildup is reduced by only $\approx 0.16\,\%$ due to a remaining dynamic frequency mismatch. Thus the statement given above is justified, that a further increase of the control loop's UGF was not necessary.

3.4. The first LSW experiment with production resonator

Production resonator linewidth. A precise value for the production resonator's *FWHM* was determined by measuring its transfer function for an AM of the incident beam to the AM of the transmitted beam. The modulation was imprinted on the beam by the AOM, which was part of the infrared laser. The result of this measurement is shown in Fig. 3.19. It was fitted with a simple one-pole lowpass filter function, which is a good approximation for this kind of transfer function of a cavity for frequencies much smaller than the free spectral range

$$TF(\nu) = \frac{1}{1 + i\frac{2\nu}{FWHM}} \quad .$$

From this fit an average value of

$$FWHM = 120.5\,\text{kHz} \pm 1.2\,\text{kHz}$$

was obtained.

Comparison of the result for *FWHM* with the free spectral range of the production cavity yields that the latter is approximately 145 times the linewidth. Comparison of this value with Fig. 3.18 demonstrates how rapidly the free-running frequency mismatch between the production cavity resonance and the incident light increased. In an unstabilized situation the complete *FSR* of the production resonator was scanned within only about 130 ms! This is most likely much faster than in most laboratory experiments involving optical resonators.

Resonant power buildup. The resonant power buildup PB_{max} of the production resonator was a crucial value to determine its circulating power and thus the sensitivity of the experiment. It was routinely determined by measuring the power incident on the input coupler and transmitted through the far end mirror. To get reliable results it was obviously important to measure the transmission of the far end mirror T_{out} very precisely. To improve precision of this measurement it was conducted in-situ, i.e. when the mirror was at the same temperature and located and aligned in the same way, as it would have been to form the production resonator. One reason for this step was that the in-situ temperature deviated by up to 10 °C from ambient conditions. The procedure was to deliberately misalign the

3. ALPS I project - Particle physics with high-power green light

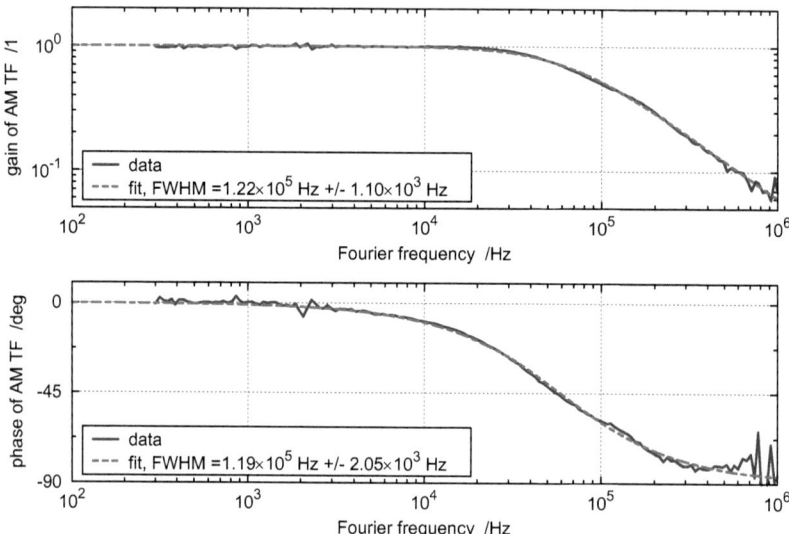

Fig. 3.19: Determination of production resonator linewidth by measuring the transfer function of an AM of the power in front of the cavity to an AM of the transmitted power. The data was fitted with a single-pole lowpass filter function to determine the cavity's linewidth.

input coupler until the cavity no longer existed. Then the combined transmission of input coupler and far end mirror was measured and then this value was divided by the known transmission of the input coupler. The latter value could be obtained with high precision at the time of installation because the mirror was always at room-temperature and its deliberate misalignment in this measurement procedure was tiny due to the long production cavity. Determination of the far end mirror transmission prior to installation and in-situ by the method just described resulted in a difference of up to a factor of two. It was decided to stick to the value from the in-situ measurement due to the big differences in the ambient conditions. This value

3.4. The first LSW experiment with production resonator

was $T_\text{out} = 170\,\text{ppm} \pm 12\,\text{ppm}$. Then from several measurements of the incident and the transmitted power a value of the resonant power buildup based of the value of T_out was obtained as

$$PB_\text{max}^{(EM)} = 44 \pm 3.8 \quad.$$

In order to test the reliability of this measurement the value of T_out was compared with the result from the *FWHM* measurement (see Fig. 3.19). Presuming that the free spectral range was known with high precision (which was the case as explained above), Eq. (2.18) could be used to derive from *FWHM* and T_in a value for the fractional round trip losses of $A_\text{p,a} + A_\text{p,s} = 2.0\,\% \pm 0.15\,\%$ corresponding to a value for the resonant power buildup derived via Eq. (2.19) of

$$PB_\text{max}^{(FWHM)} = 49 \pm 5 \quad.$$

The achieved value of $PB_\text{max}^{(FWHM)}$ was obviously limited by the high fractional round trip losses due to scattering and absorption $A_\text{p,a} + A_\text{p,s}$. These were expected to originate mainly from the reflection of the eight facets of the vacuum tube windows that had to be traversed per round trip. From the measurements above their reflection per surface was estimated to be $< 0.25\,\%$, which is in acceptable agreement with the assumptions made in the design phase of the experiment. Removal of these windows from the light path inside the production resonator should enable one to increase the resonant power buildup and thus the sensitivity of the LSW experiment significantly. Throughout the ALPS I experiment $PB_\text{max} = PB_\text{max}^{(EM)}$ was applied because the light, which is transmitted through the far end mirror is a direct copy of the circulating power, which in turn is one of the key parameters when calculating the sensitivity of an LSW experiment.

Power buildup measurement methods. As described in the previous paragraphs, the resonant power buildup was measured by two different techniques. The first one, which resulted in $PB_\text{max}^{(EM)}$ was sensitive to RMS deviations of the circulating power from its maximum because the power incident on and transmitted through the cavity were measured with an averaging power meter. In contrast, the method to determine $PB_\text{max}^{(FWHM)}$ was mostly insensitive to those effects. The upper graph

3. ALPS I project - Particle physics with high-power green light

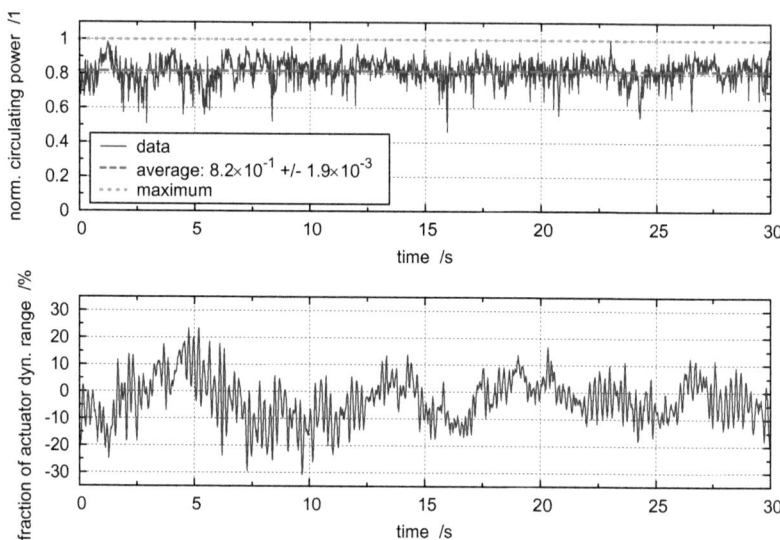

Fig. 3.20: Typical time series of the normalized circulating power (top) and of the actuator signal (bottom) in a stabilized state. The latter was calibrated to show the fraction of its range, which was necessary to maintain the resonant state.

of Fig. 3.20 shows a typical time series of the circulating power inside the production resonator together with its maximum and average value. From this measurement a difference between average and maximum of 18 % was obtained, which was in agreement with the error bounds of the power buildup measurements.

In the context of Fig. 3.18 it was shown, that with active frequency stabilization the frequency mismatch between incident light and cavity resonance frequency was reduced below significance and thus was not expected to cause fluctuating circulating power. Incident power noise was usually on the order of only a few percent and its influence would have been sensed in front of and behind the cavity and would

3.4. The first LSW experiment with production resonator

thus mostly cancel out of the result for $PB_{max}^{(EM)}$. Therefore the reason was most likely due to an imperfect overlap of the incident mode and the eigenmode of the production resonator, i.e. an $\eta_{00} < 1$ in Eq. (2.13). Such an imperfect overlap can be due to stationary reasons, like the slight distortion of the incident SHG beam profile. It can also be due to dynamic reasons, namely fluctuating alignment caused by vibrational or acoustic noise, also called pointing.

Nonetheless the upper graph in Fig. 3.20 also shows that despite of the fluctuations the average value of the circulating power was well defined with small uncertainty, which allowed to reliably define a sensitivity of the overall LSW experiment.

Range of the actuator. A somewhat more technical issue is answered by Fig. 3.20, too. Its lower graph shows the corresponding fraction of the actuator range, which was necessary to maintain the resonant state, i.e. to maintain a high power buildup. If a realistic search for WISPs was conducted, it was important to achieve a high duty cycle of the stabilized system because otherwise the time efficiency of the LSW experiment would have been seriously degraded and the advantage of the implementation of an optical resonator could have turned into a disadvantage. After each loss of the stabilized state of the system the automatic lock acquisition electronics put the system back into its stabilized state within approximately 1 s. Thus it was desirable to maintain the stabilized state after each lock acquisition for at least 30 s to be able to reach a duty cycle $> 96\,\%$. The lower graph of Fig. 3.20 indicates that this could be achieved. In general, considerably longer times of permanent stablization were achieved. A typical duty cycle of the system during a WISP search was around or even above 95 %. The dominant fraction of the actuator signal in Fig. 3.20 was contributed by a rather sinusoidal signal at frequencies between 6 Hz and 9 Hz. This was in agreement with the dominant free-running frequency noise structure visible at the same frequencies in Fig. 3.17.

Although the frequency mismatch between cavity resonance and incident light was changing rapidly, the acquisition of the stabilized state did not cause problems.

3. ALPS I project - Particle physics with high-power green light

Circulating power during WISP search

In summary, the experiments discussed in this section could clearly show that it is technically possible to incorporate an 8.6 m long linear production resonator into a large-scale LSW experiment located at a site, which was not specifically selected or designed for this and correspondingly rather noisy. With an average power buildup of 43 an exemplary search for WISPs was conducted, which in sum collected data within 29 data frames of the single-photon detector, each with a duration of 1 h. These frames were distributed over the various experimental parameter sets necessary for the different WISP species of interest (see 3.2. The optical power circulating inside the production resonator is shown in Fig. 3.21. As described above it was obtained from measurements of the incident and reflected power, which were continuously recorded together with the ambient temperatures at the cavity mirror locations and the magnetic field strength over the whole run. Before and after each data frame the resonant power buildup $PB_{max}^{(EM)}$ was determined. The alignment was always good enough to assure that the circulating power could be maintained with a duty cycle of 95 % and higher. Obviously the highest instantaneous circulating power recorded over the whole measurement run was 38.8 W corresponding to a maximum power buildup of 49, which is in excellent agreement with the value of $PB_{max}^{(FWHM)}$ obtained above. To accomplish this WISP search the production resonator had to remain at comparable performance levels for at least half a month. This was due to the complexity of the overall LSW experiment and measurement procedure (the latter permanently involving between three and five people), and it was due to the necessity to record additional data to characterize the apparatus and make the results reliable.

To the best of the author's knowledge the ALPS I (phase 1) experiment was the first large-scale LSW experiment reported in literature, which was able to search for WISPs with a production resonator. Although originally meant only as a demonstration for the compatibility of a large-scale LSW experiment with a long production resonator, the experiment obtained exclusion limits, which already came close to the best reported so far. These results were published in [44]. In this publication also a

3.5. ALPS I (phase 2) - the world's most sensitive WISP detector

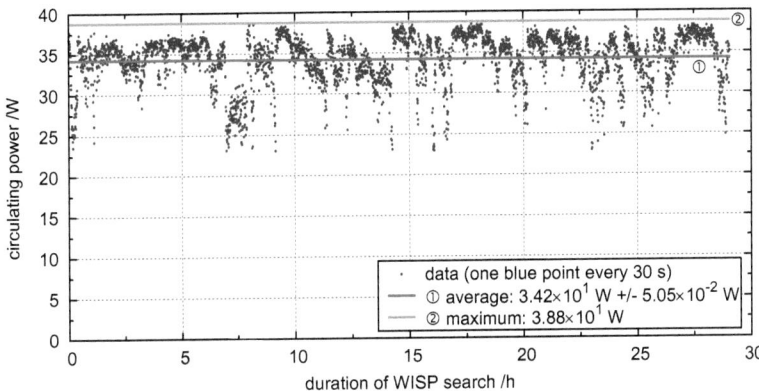

Fig. 3.21: Circulating power inside production resonator during exemplary WISP search measurement run of the ALPS I (phase 1) experiment. Out of this data 25 valid 1 h-data frames were selected, which were not influenced by cosmic radiation. The whole measurement run lasted for approximately half a month.

detailed description can be found of the overall LSW experiment, of the exemplary measurement run and the evaluation of its data.

3.5. ALPS I (phase 2) - the world's most sensitive WISP detector

In section 3.4 it was argued that the incorporation of a production resonator into a large-scale LSW experiment is beneficial if its sensitivity should be increased by significant enlargement of the number of primary particles, i.e. the number of photons in the production region. Such an LSW experiment was set up and it was demonstrated that the implementation of a long production resonator is compatible

3. ALPS I project - Particle physics with high-power green light

with the general requirements of the overall experiment.

Consequently, now its sensitivity for the various WISP species should be increased as much as possible. The realization of this aim benefited heavily from the precise characterization of the precursor experiment presented in the previous section, especially from the identification and selection of the important noise sources and from the tests of the long-term performance. In the following the upgrade process is described. Unless otherwise noted all described devices and experiments were developed or conducted respectively by the author of this thesis.

3.5.1. Experimental setup of ALPS I (phase 2) experiment

To improve the sensitivity as much as possible the following steps were taken:

- It was discussed in the context of ALPS I (phase 1) that the setup of its production resonator did not result in lock acquisition problems. Consequently, the resonant power buildup of the production resonator for ALPS I (phase 2) was planned to be increased considerably by a strong reduction of the round trip losses and consecutive selection of an input coupler with much smaller transmission.

- The laser power incident on the production resonator was planned to be increased by substitution of the single-pass SHG stage by a resonant SHG stage.

- The detection stage was equipped with an improved single-photon detector of type PIXIS 1024B by Princeton Instruments. It was operated with the same parameters as the original one but its quantum efficiency was 96 % and its dark noise was as low as 46 $\frac{\text{photons}}{\text{1h-frame}}$ following approximately a poissonian distribution [114, 88]. Selection, characterization and operation of this detector were not part of this thesis and thus it is not considered here further.

The improved experimental setup of the injection stage and the production resonator is depicted in Fig. 3.22. Apart from this the overall setup of the LSW

3.5. ALPS I (phase 2) - the world's most sensitive WISP detector

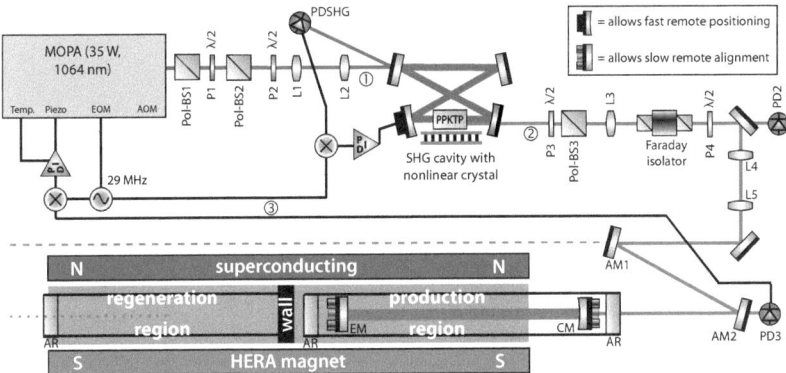

Fig. 3.22: Detailed schematic view of the improved optical setup of the injection stage and the production resonator of ALPS I (phase 2). Red lines (1) depict a laser wavelength of 1064 nm and green ones (2) of 532 nm. Black lines (3) depict electric signals.

experiment did not change significantly compared to Fig. 3.12. The changes are explained in the following.

SHG resonator

The former single-pass SHG stage was rather limited in its output power and therefore substituted by a resonant one. As nonlinear material PPKTP was used. The SHG resonator was identical to the device described in 2.3 and thus its design and experimental setup is not repeated here. Again the light power incident on the production cavity was measured with photodetector PD2. The infrared power reflected from the SHG resonator was detected with PDSHG.

From this resonant SHG stage an output power of 5 W was available compared to the 900 mW obtained from the single-pass SHG stage of ALPS I (phase 1).

153

3. ALPS I project - Particle physics with high-power green light

Production resonator

For the same reasons that were enumerated in the context of the ALPS I (phase 1) experiment the green light produced by the resonant SHG was guided through a variable attenuator, a Faraday isolator, beam-shaping lenses, a $\lambda/2$ waveplate and alignment mirrors to the production resonator. But in comparison with the precursor experiment this time photodetector PD3, which measured the power reflected from the cavity, was located closer to the cavity and separated from the input coupler CM only by a single mirror, which was hit by the reflected beam under an angle as small as possible. This setup change strongly reduced polarization dependent effects in the output signal of PD3 and eased the commissioning of the production resonator.

The resonant power buildup factor of the ALPS I (phase 1) production resonator was shown to be limited mainly by the round trip losses introduced by the windows of the production vacuum tube. In order to increase PB_{max} the production vacuum tube of the ALPS I (phase 2) experiment was extended to encompass both resonator mirrors. Thus the production resonator was now an ideally empty resonator, whose resonant power buildup was limited by the losses introduced by its mirrors only. These losses were expected to be especially low if one could manage to keep the mirrors during their installation phase as clean as possible. This was easy to arrange for the input coupler of the production cavity, which was installed into a vacuum tank located on the optical table of the injection stage. From the beginning of the ALPS I experiments the whole optical table of the injection stage was located underneath a so-called flow-box equipped with high efficiency particulate air (HEPA) filters. They produced a steady flux of very clean air to fill the whole area of the injection stage. Measurements with a particle counter resulted in values of ca. 100 particles/ft^3 while people are moving in the area and even less otherwise, which was considered acceptable.

In contrast, for the installation of the far end mirror one had to insert it together with its mount into the opened production vacuum tube from in front of the magnet outside of the laser hut. There, all parts were exposed to the unfiltered air of building 55, which was measured to consist of approximately 5×10^5 particles/ft^3.

3.5. ALPS I (phase 2) - the world's most sensitive WISP detector

Consequently, it was very likely that this exposition would lead to increased losses of the far end mirror. Therefore an auxiliary clean-room tent was constructed to cover the area in front of the magnet and thus to keep the far end mirror clean. It consisted of a large cleaned foil and a movable ventilation system, which was equipped with a HEPA filter stage and which blew air into the tent from above. It worked rather efficiently reducing the particle density measured with a particle counter to values comparable to those measured on the optical table of the injection stage underneath the flow-box. Thus both resonator mirrors could be installed inside of the production vacuum system under clean conditions.

Coating and planarity of the resonator mirrors were the same as for ALPS I (phase 1) and hence the same losses per mirror of approximately 0.14 % were assumed. In order to maximize the resonant power buildup, again the construction of an impedance matched resonator was desired, leading to a design value for the transmission of the input coupler of 0.28 %. Resonator length and free aperture hardly changed and thus the mirrors had the same radii of curvature as for ALPS I (phase 1). Precise measurements of the transmission of the input coupler prior to installation resulted in a value of 0.16 % ± 0.01 %, which was accepted because the resonant power buildup depends only weakly on the transmission of the input coupler if the cavity is at least nearly impedance matched.

To start with a new and clean set of resonator mirrors also the far end mirror was exchanged. After installation of both resonator mirrors the production vacuum system was closed and evacuated until the rest gas pressure was smaller than 10^{-5} mbar. Then the transmission of the new far end mirror was determined by the same in-situ measurement procedure extensively described in the previous section. The obtained value was 88 ppm ± 5 ppm.

For some experiments a considerably higher pressure of 0.1 − 0.2 mbar could be created by filling purified argon gas into the vacuum system. This was done to 'fill' the sensitivity zeroes in the parameter space plot at WISP masses above 1 meV (see Eq. (3.1)).

3. ALPS I project - Particle physics with high-power green light

Control schemes

The control scheme of the ALPS I (phase 2) production cavity was identical to that of the ALPS I (phase 1) experiment with one exceptions. The large reduction of the production resonator linewidth caused the remaining mismatch between laser frequency and production cavity resonance frequency in a stabilized state to degrade the stabilized power buildup. In order to mitigate this, the control loop electronics were enhanced by a broad notch filter centered around 200 kHz, which rudimentary suppressed the whole complex resonance structure observable in the ALPS I (phase 1) measurement of the control loop's open-loop gain in Fig. 3.16. This step allowed the enhancement of the control loop's UGF from 30 kHz to 51 kHz. The above mentioned degradation of the stabilized power buildup was reduced by this step to an acceptable amount.

Furthermore an additional control loop had to be implemented for the resonant SHG stage, which kept the SHG cavity resonant with the incident NIR light. The control loop itself actuated on the SHG resonator length and was a copy of that one described in 2.3. But in the case of ALPS I (phase 2) the SHG control loop was coupled to the outer control loop, which minimized the frequency mismatch between laser and production cavity resonance. Here the term 'coupled' is used instead of 'nested' as the inner loop was not part of the electric signal path of the outer loop and thus did not impose any significant linewidth limitation on the outer loop. If at any given point in time the outer loop compensated for some frequency mismatch, it did that by actuating on the NIR laser frequency, and thus the resonator length of the SHG stage had to be changed correspondingly and fast enough by the inner loop such that its resonance followed the NIR laser frequency change. In such a case control theory finds that ideally the inner loop should have a higher UGF than the outer loop [2]. Unfortunately, this was not possible in the ALPS I (phase 2) setup because on the one hand the outer loop needed an as high UGF as possible for the reasons just explained. On the other hand the inner control loop UGF was restricted to approximately 10 kHz due to complex mechanical resonance structures starting from approximately 20 kHz. These structures originated from the necessity

3.5. ALPS I (phase 2) - the world's most sensitive WISP detector

to move a heavy one inch optic at high speed. Despite this problem the coupled control loops worked in the case of ALPS I (phase 2), which was due to the compact and low finesse design of the SHG resonator. Its optical round trip length was only 25 cm corresponding to a large $FSR = 1.2\,\text{GHz}$. As normal for a high power resonant SHG stage its finesse was low, around a value of 40, which led to a huge linewidth of $FWHM = 30\,\text{MHz}$. The fast outer loop had to compensate steadily for the free-running frequency mismatch between laser and the resonance of the production resonator. The inner loop UGF was lower and thus could not compensate for that RMS mismatch, which was acquired within observation times on the order of the inner loop's UGF of 10 kHz. Comparison with Fig. 3.18 gives a value of this remaining frequency mismatch between NIR laser and SHG cavity resonance of less than 10 kHz. If this value is inserted into Eq. (2.19) the corresponding drop in the converted power can be calculated to be completely negligible. Here it was assumed that the SHG resonator length fluctuated much less than the production cavity length, which was estimated to be true.

The error signals of the SHG resonator were obtained with photo detector PDSHG via the PDH technique from the same modulation sidebands at 29 MHz, which were also used for the production cavity stabilization. The cavity linewidth of the SHG resonator was broad enough to transmit a considerable fraction of the modulation sidebands to the production cavity. Hence both control loops obtained big enough error signals.

3.5.2. Results of ALPS I (phase 2) and discussion

SHG resonator

The resonant PPKTP SHG stage for the ALPS I (phase 2) experiment had to work reliably more or less continuously for several months and thus it was operated with some safety margin with respect to the maximum output power obtained in 2.3 to avoid crystal damage and to make the frequency stabilization control loop more robust. After settling of its output power as described in 2.3 it was set to 5 W at a wavelength of 532 nm while approximately 10 W at a wavelength of 1064 nm were

157

3. ALPS I project - Particle physics with high-power green light

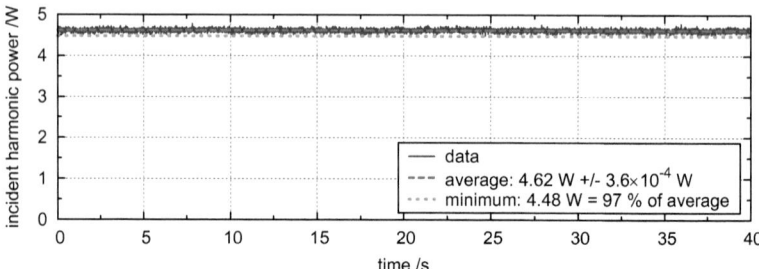

Fig. 3.23: Typical time series of the harmonic power incident on the input coupler of the production resonator.

incident. Approximately 4.6 W were constantly available at the input coupler of the production resonator. The output power showed a good stability with fluctuations typically on the order of only 3 % as is shown in Fig. 3.23. The lateral modeshape was close to gaussian without the distortions reported in the previous section in the context of the single-pass SHG stage. The TEM_{00} resonance frequency of the SHG resonator was stabilized to the incident laser frequency by actuation on its roundtrip length.

Production resonator

Production resonator linewidth. In any case the extension of the vacuum system to encompass both resonator mirrors increased the vacuum distance on which massive hidden photons (MHPs) could be produced from 6.3 m to 8.6 m, which directly increased the sensitivity to this WISP species (see Eq. (3.1)). Furthermore the elimination of the vacuum tube windows from the production cavity and the efforts for increased cleanliness led to a reduction of the cavity linewidth and thus to an increase of the power buildup factor by nearly one order of magnitude. Fig. 3.24 shows a measurement of the linewidth obtained by the same method described in the previous section.

3.5. ALPS I (phase 2) - the world's most sensitive WISP detector

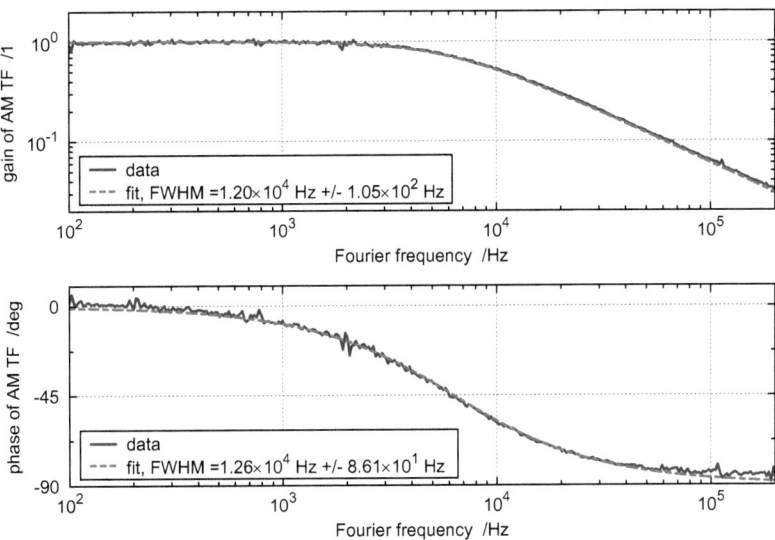

Fig. 3.24: Determination of the linewidth of the ALPS I (phase 2) production resonator, measured with the same technique described in the context of Fig. 3.19.

From a fit to this measurement a value of

$$FWHM \;=\; 12.3\,\text{kHz} \;\pm\; 68\,\text{Hz}$$

is obtained corresponding to passive fractional losses introduced by the resonator mirrors of $A_{p,a} + A_{p,s} = 0.275\,\% \pm 0.01\,\%$ (see Eq. (2.18)), which was in excellent agreement with the value assumed in the design phase. Combination of these losses with the measured mirror transmissions implied a value for the resonant power buildup derived via Eq. (2.19) of

$$PB_{\max}^{(FWHM)} \;=\; 325 \;\pm\; 16 \quad.$$

3. ALPS I project - Particle physics with high-power green light

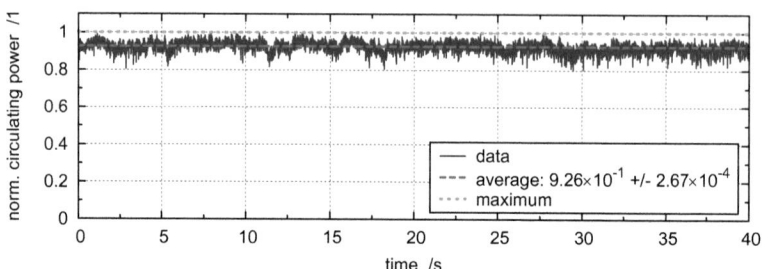

Fig. 3.25: Typical time series of the circulating power in a stabilized state of the ALPS I (phase 2) experiment.

Power buildup measurement methods. As also described in the previous section another value for the resonant power buildup could be obtained from a measurement of the light transmitted through the far end mirror. In the ALPS I (phase 2) experiment the discrepancy of $PB_{max}^{(EM)}$ from $PB_{max}^{(FWHM)}$ was on the order of 10%, which was slightly higher than the difference between maximum and average value of a typical time series, which is shown in Fig. 3.25. Correspondingly, during this characterization the circulating power inside the production resonator was always above 1 kW!

In the ALPS I (phase 1) experiment it turned out that remaining frequency noise in the stabilized state was not an issue and most of the discrepancy between the two methods to determine the power buildup was due to alignment fluctuations. This was different in ALPS I (phase 2) because the linewidth of its production cavity was smaller by an order of magnitude, while the ambient noise did not change. In analogy to Fig. 3.18, Fig. 3.26 shows the RMS value of the frequency mismatch between laser and cavity resonance in the present case. As was explained above the control loop's UGF was increased to 51 kHz but still the stabilized RMS mismatch remained at 13.6 % of the cavity's *FWHM* for observation times of seconds. Via Eq. (2.19) this value could be translated into an RMS power buildup degradation of 6.9 %.

3.5. ALPS I (phase 2) - the world's most sensitive WISP detector

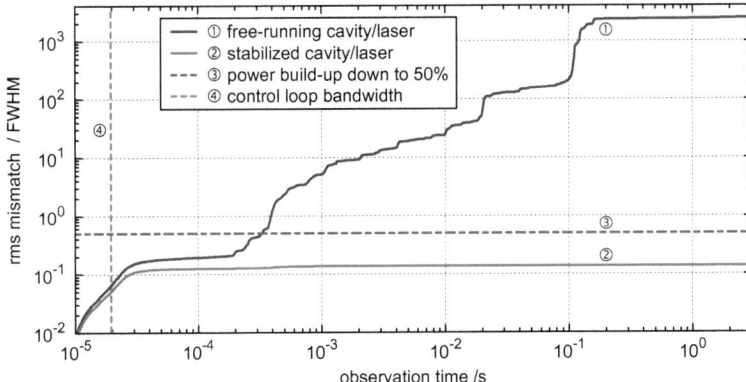

Fig. 3.26: The ALPS I (phase 2) RMS mismatch between laser frequency and production cavity resonance frequency in an unstabilized (so-called free-running) and in a stabilized state is shown above the length of the observation time. As guides for the eye the UGF of the frequency control loop and the RMS mismatch corresponding to 50 % power buildup loss are marked, too.

Thus only ca. 3 % of the power buildup discrepancy stated above had to be attributed to pointing or a mismatch between incident mode and eigenmode. This was an improvement compared to ALPS I (phase 1) and most likely due to a combination of the following changes made during the upgrade to ALPS I (phase 2):

- There was no doughnut-like distortion of the incident beam visible (originally caused by single-pass SHG stage in ALPS I (phase 1)).

- Air currents and acoustics could no longer dynamically change the alignment of the production resonator eigenmode relative to the incident beam because both resonator mirrors were now located inside the vacuum system.

3. ALPS I project - Particle physics with high-power green light

- If such a dynamic alignment change was caused by vibrational noise it should have been suppressed in ALPS I (phase 2) because the rigid vacuum tubes somewhat fixed the alignment of the input coupler relative to the far end mirror.

Mirror damage. As was explained above up to this point only ca. 10 W of the 35 W supplied by the infrared laser were used to generate green light as input for the production cavity. The ALPS I (phase 2) experiment aimed at a maximized sensitivity and hence it might seem questionable, why no efforts were made to further increase the incident harmonic power.

The reason was that already at the current level of circulating power of about 1 kW the production resonator mirror coatings had a very limited lifetime of approximately $20 - 30$ h. After this period of time the resonant power buildup of the cavity suddenly started dropping and was reduced below one third of its original value within only about two additional hours. This effect occurred repeatedly for at least four times although huge efforts were made to assure clean mirror facets, like precise visual inspection of the mirror facets prior to installation, cleaning of the vacuum system or further reduction of the particle density in the air during mirror installation. Due to this problem several mirror sets had to be used during the characterization and WISP search phases of the ALPS I (phase 2) experiment. However, they did not differ too much in their performance.

Visual inspection of the damaged mirror revealed a darkened region at the place where the beam of the eigenmode hit the mirror. It should be noted here that the intensity on the mirror facets was always far below their typical laser-induced damage thresholds stated by the supplier to be on the order of $2\,\frac{\text{MW}}{\text{cm}^2}$ for CW radiation. In the context of ALPS I (phase 1) it was explained that the radius of the beam on the mirrors was $1-2$ mm. At a circulating power of 1.5 kW the mirrors had to withstand a peak intensity not bigger than $100\,\frac{\text{kW}}{\text{cm}^2}$.

Due to its comparatively small radius on the mirrors the eigenmode of the plane-

3.5. ALPS I (phase 2) - the world's most sensitive WISP detector

concave resonator could be shifted up or down by alignment of both mirrors until it reached a region, which was not hit by the beam before. This procedure was done while the mirrors were still in vacuum. After repositioning of the eigenmode the resonant power buildup was up again, often it nearly reached its original value but then dropped again after another approximately 10 hours. This demonstrated that the power buildup reduction was not caused by some material, which covered the mirror facet and reduced its reflectivity more or less equally. Moreover, no obviously higher damage probability for one or the other resonator mirror was observed although the intensities differed by a factor of two. Finally, it made no obvious difference for the mirror lifetime if the cavity was operated at gas pressures below 10^{-5} mbar or 0.2 mbar. However, the thermal load on the mirrors appeared to vary considerably, which was estimated from the rather different strength of thermal alignment drifts at the two gas pressures.

The ALPS I (phase 2) experiment was thus limited by the lifetime of the chosen mirrors and a further enhancement of the circulating power did not seem to be meaningful.

Circulating power during WISP search

Finally, after implementation of all improvements and upgrades to the original experiment a search for WISPs was conducted with the ALPS I (phase 2) experiment. Two of the three main sources for the significantly improved sensitivity of the experiment were

- the implementation of a high power laser source for light at a wavelength of 532 nm providing nearly 5 W of optical power to the production resonator,

- and the improvement of the power buildup factor of the production resonator by nearly a factor of 7, requiring a cavity with a linewidth of only 12 kHz and a finesse of 1400.

The performance of these devices during the whole WISP search is shown in Fig. 3.27. As already explained in the context of ALPS I (phase 1) the circulating power was obtained from measurements of the incident and the reflected power. The power buildup was calculated as ratio of circulating power divided by incident power. All in all 55 data frames, each of 1 h duration, were recorded, distributed over a time period of 3 months. In the same manner as in ALPS I (phase 1) a duty cycle of 95 % or higher was achieved when the optical system was aligned well enough.

The overall data taking period was considerably longer than in ALPS I (phase 1) because, first, more data was collected, second, the cavity mirrors had a limited lifetime, and third, problems occurred with the superconducting magnet, which repeatedly underwent so-called quenches. In a quench the magnet, which was constantly driven with a current of 6000 A, spontaneously lost its superconducting state and rapidly heated up causing all the cooling helium to evaporate nearly instantly. On the one hand such a process induced massive vibrations and distortions into the optical system, leading to realignment efforts of varying complexity before it could be stabilized again. On the other hand such quenches, if they occur repeatedly, tend to damage the magnet reducing its maximum magnetic field. To reduce this risk, time consuming efforts to investigate the reasons of such quenches were undertaken.

The stretches in time, in which the performance of the production resonator was influenced by these quenches, is depicted in Fig. 3.27 by vertical yellow colored bars. After the third quench time-consuming realignment of the resonator eigenmode with respect to its free aperture was necessary, which was postponed to the end of that day and caused the following data frame to show rather bad performance.

Beyond the production resonator the resonant SHG stage had to be stabilized also as part of two coupled control loops. Its performance is shown in the same figure. The steps in the provided power to the production resonator were without exception due to changes of the working point. Obviously its implementation did not have any negative influence on the overall performance.

The performance of the production resonator can be better judged if the influence of the varying incident power is removed from the circulating power, which was done in the third part of Fig. 3.27 where the effective power buildup of the production cavity is plotted as ratio of circulating power divided by incident power.

3.5. ALPS I (phase 2) - the world's most sensitive WISP detector

The high circulating power repeatedly caused the cavity mirrors to degrade after $20-30\,\mathrm{h}$. In Fig. 3.27 the stretches of time, during which the cavity was operated with a certain mirror set, are marked by the red and green horizontal bars. The rather slow drop of the power buildup at the end of the region marked by the red bar (compared with the fast ones caused by quenches or general losses of the stabilized state) indicates the degradation of the mirror set, which was used from the beginning of this WISP search. After installation of a new set the power buildup was up again but a bit lower than before due to slightly higher losses. Together with the installation of this set purified argon was injected into the production vacuum tube at a pressure of $0.1-0.2\,\mathrm{mbar}$ to collect data for 'filling' the sensitivity zeroes in the parameter space plot. Those times, at which data was recorded with argon gas in the production region are marked with violet bars in Fig. 3.27. Maybe other materials contained in this gas caused the slightly higher losses of the second mirror set. The second mirror set started to degrade, too, right at the end of the measurement period shown here.

The average circulating power, excluding such time periods influenced by quenches (and also the bad performance period after quench number three), amounted to 1.2 kW, while the first mirror set was installed. After installation of the second set the average value dropped but still kept at nearly 1 kW. The average power buildup varied between 230 and 300 depending on mirror set and time span. An interesting feature was its rise to its maximum average at times between 14 h and 21 h. This maximum occurred right after a realignment and repositioning of the eigenmode with respect to the magnet's free aperture and also with respect to the point where the eigenmode hit the mirrors. Therefore it is likely that the round trip losses introduced by the mirrors were not equal for varying positions on the mirror facets.

Finally, variations of the circulating power on the order of 20 % were not that important for the sensitivity of the experiment because the coupling constant of the quantum fields of light and WISPs scales with $\sqrt[4]{P_\mathrm{circ}}$ (see Eq. (3.1)).

3. ALPS I project - Particle physics with high-power green light

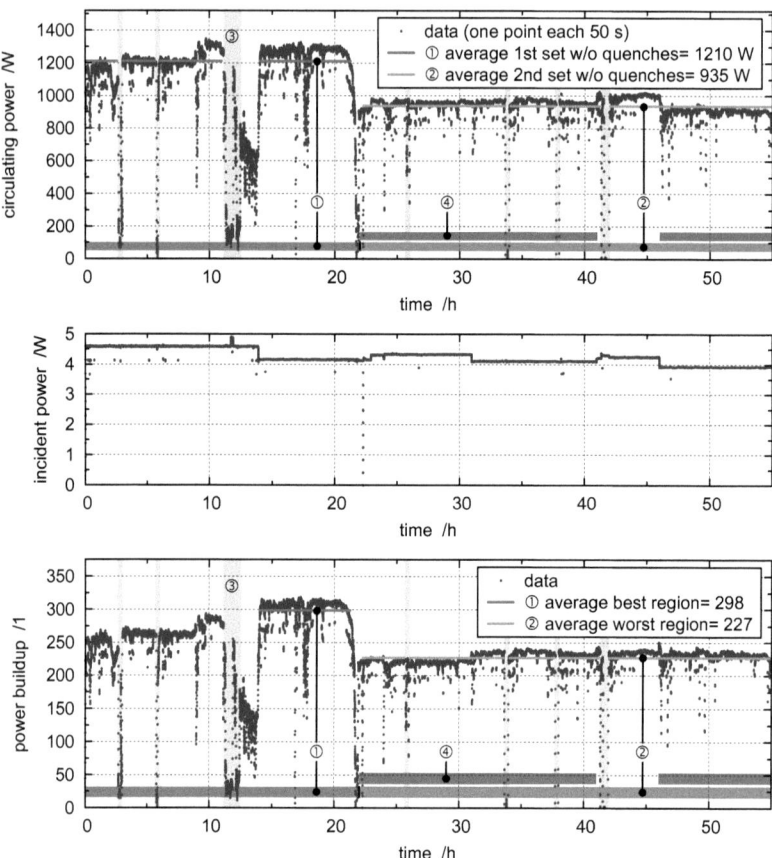

Fig. 3.27: Circulating and incident power and derived power buildup of the ALPS I (phase 2) production cavity during WISP search. Yellow regions (3) were influenced by quenches. Red (1) and green bars (2) at the bottom depict cavity mirror sets, violet bars (4) depict measurements with argon gas. Regions not covered by average lines were excluded from best/worst average calculation. Overall average circulating power was 1.04 kW.

3.5. ALPS I (phase 2) - the world's most sensitive WISP detector

Most sensitive experiment in the world

The recorded data could be translated into upper limits for the value of the coupling constants, describing the linear coupling between the quantum fields of light and a WISP species. To the best of the author's knowledge the obtained results of the ALPS I (phase 2) experiment for APs and MCPs are the most stringent laboratory upper limits among all LSW experiments in the world for masses up to at least 10 meV. For MHPs the mass range of most stringent results extends from approximately 0.15 meV to at least 10 meV with some tiny regions in between, where the LIPSS experiment achieved a higher sensitivity. The results are depicted in Fig. 3.28 as those labeled 'ALPS' and 'ALPS(gas)'. This figure is a compilation of those published in [45]. The evaluation of the WISP search data itself was not part of this thesis.

For APs and MCPs there are stricter upper limits from astrophysical observations, however these results rely on inherently uncertain models of the processes involved. In any case there are no more stringent bounds for MHPs with masses from 0.15 meV to 3 meV than those obtained by ALPS I.

In the context of 3.2.2 it was explained that MHPs might have formed a hidden microwave background, which contributes to the effective number of neutrino species [67]. The corresponding parameter space of MHPs that could explain these recent results from [70] is shown in Fig. 3.28 in the bottom-left graph as reddish shaded region. The ALPS I (phase 2) experiment could restrict this model to a very small remaining range in parameter space.

To the best of the author's knowledge the ALPS experiment also was the first LSW experiment reported in literature, which utilized a background gas in the production and regeneration region to 'fill' the sensitivity zeroes in the parameter space plots at masses starting from approximately 1 meV. The exclusion limits obtained from this gas enhancement are also shown in Fig. 3.28 and labeled 'ALPS(gas)'. The overall experiment of ALPS I (phase 2), its complete results and data evaluation procedure were published in [45]. The importance of this publication for the field of research was stressed by the publication of a referencing note in the *Research*

3. ALPS I project - Particle physics with high-power green light

Highlights section of *Nature* [57]. Additional information on the optical part of the experiment was published in [97].

The costs of the overall experiments including cooling of the magnet and purchase of all components but excluding staff costs amounted to approximately half a million euro [88]. From the very first discussions about possibilities for ALPS I (phase 1) to the publication of these final results of ALPS I (phase 2) it took only two years.

3.5. ALPS I (phase 2) - the world's most sensitive WISP detector

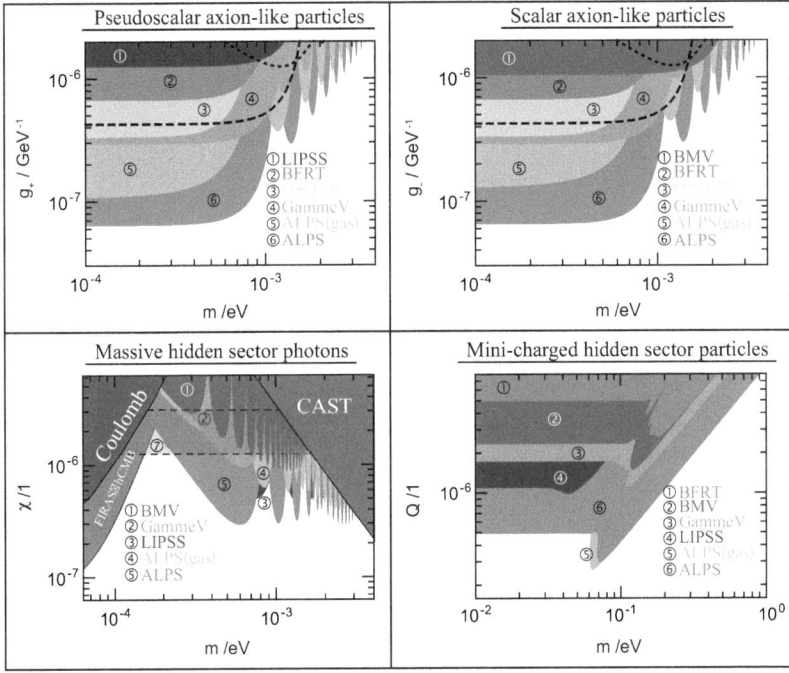

Fig. 3.28: Exclusion limits (95 % confidence intervals) for relevant WISP species set by the ALPS I (phase 2) experiment (denoted 'ALPS' and 'ALPS(gas)') and by others (taken from [45]). With the exception of small parameter regions of MHPs the ALPS exclusion limits are the world's most stringent limits obtained from laboratory experiments in the meV-region. For MHPs of masses $0.15 - 3\,\text{meV}$ they are even the most stringent in general. In the case of APs the broken/dotted lines show limits derived from searches for magnetic vacuum birefringence and dichroism [156]. The validity of a currently proposed model [67], which utilizes MHPs to explain the deviation of the effective number of neutrino species from its SM value (as suggested by recent WMAP data) could be constraint to a tiny parameter region (reddish-shaded region (7) in bottom-left plot).

3. ALPS I project - Particle physics with high-power green light

As was explained above the exact parameters of the LSW experiment varied to a certain extend. However, for rough comparison Tbl. 3.2 lists the approximate values of all parameters, which were relevant for the sensitivity of ALPS I (phase 2).

parameter	value
B	5.0 T
L_m	4.3 m
λ_0	532 nm
P_{inc}	4.6 W
PB	275
n-1	0 and 5×10^{-8}
detection efficiency	82 %
detection frame duration	3600 s
average dark counts per frame	46
confidence interval	95 %

Table 3.2: Approximate parameters of the ALPS I (phase 2) experiment, which resulted in the exclusion limits shown in Fig. 3.28.

3.6. Summary and outlook

The ALPS collaboration operates a large-scale LSW experiment to search for WISPs. The ALPS I experiment was the first LSW experiment worldwide, which utilized an optical resonator comprising the production region to enhance the amount of primary particles, namely photons. Application of this technique offered the chance to circumvent sensitivity limitations set by the average power levels of available pulsed lasers, which up to now have been the basis of all other LSW experiments in the world. The author of this thesis developed, commissioned, characterized and maintained the optical injection stages with their second harmonic generators, the production resonator and the necessary control loops.

3.6. Summary and outlook

In this chapter, first, the various types of WISPs were introduced, which the ALPS I experiment was sensitive for. Evidence for them was collected from physical observations, which are hardly understood in the scope of the Standard Model of particle physics and which might be related to or better explained by the existence of these types of WISPs (Sections 3.1 and 3.2).

Then the state of the art of large-scale LSW experiments in the world was summarized (Section 3.3). In Section 3.4 the challenges of the design, implementation, characterization and maintenance of the injection stage and production resonator of the ALPS I (phase 1) experiment were presented. It was explained how the author accomplished the compatibility of an 8.6 m long optical resonator with an existing large-scale LSW experiment. This compatibility demonstration included the acquisition and long-term retention of a resonant state of the production cavity, starting from an unstabilized situation, in which the frequency mismatch between cavity resonance and incident light grew from zero to the cavity's free spectral range within only 130 ms.

Finally, the ALPS I (phase 2) experiment was presented, which was an upgrade of the previous phase and which heavily benefited from the insights obtained in the precursor. It employed a high-power resonant SHG stage as source for 532 nm laser light and stored an optical power of about 1 kW in its production resonator over an integrated data taking duration of 55 h (Section 3.5).

Thanks to the incorporation of the SHG and production resonators the ALPS I (phase 2) experiment could set the most stringent upper limits obtained from laboratory experiments on the WISP types of interest! It is still the most sensitive experiment in the world.

Based on the experiences from both phases of ALPS I, the ALPS collaboration decided to plan a follow-up experiment with a by orders of magnitude higher sensitivity. By now, this experiment is in its design phase and first technical tests are under way. It will be discussed further in Chapter 4.

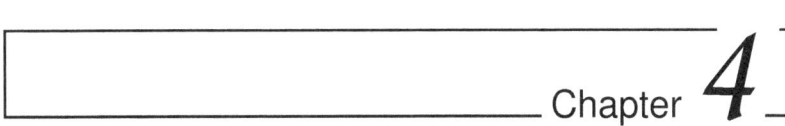

ALPS II - High-precision optical metrology boosts sensitivity

As was explained in the previous chapter, evidence for the existence of several WISP types is cumulating and therefore a finding might be just around the corner. But there are also several models, which predict WISPs with coupling constants to photons, which are still way off the sensitivity of current LSW experiments.

The implementation of a second large-scale resonator denoted as regeneration cavity, and the use of a local oscillator, allow in principle to considerably increase the sensitivity of an LSW experiment. Hence the ALPS collaboration decided to plan and build a new ALPS II experiment, which should include the implementation of such a regeneration cavity.

In Section 4.1 of this chapter sensitivity enhancement techniques are explained, which can be employed on the regeneration side of an LSW experiment. One example is the regeneration cavity. Then, in Section 4.2, a basic design study of some important aspects of the ALPS II experiment is presented. Based on this, Section 4.3 presents a possible realistic experimental setup of the optical part of a regeneration cavity enhanced LSW experiment. Finally, Section 4.4 shows the projected sensitivity of such a setup.

Some basic aspects of the proposed experimental setup have already been published in [97].

4.1. Signal enhancement on the regeneration side

On the production side of the LSW experiment, primary laser and production cavity provide an as high circulating field as technically possible. Once this is done, from an optics point of view no other techniques are known enhance the amplitude of the produced WISP field. In the previous chapter the regeneration side of the experiment was optimized only by utilization of a single-photon detector with an as low as possible dark detection rate. In order to improve the sensitivity even further, two techniques are discussed in this section, which can be implemented on the regeneration side.

4.1.1. Local oscillator

As is well known from optical metrology, the application of a local oscillator allows to overcome the electronic noise of a detector. This is an option also in the case of LSW experiments. If one detects a weak regenerated field E_r with the help of a large local oscillator (LO) field E_{LO}, then the electric field seen by the detector consists of two contributions

$$E_{det} = E_r(\omega_r, \phi_r) + E_{LO}(\omega_{LO}, \phi_{LO}) \quad .$$

Here the signal and LO field are allowed to have different frequencies ω and slowly varying phases ϕ. The detector usually measures the optical power $P_{det} \propto |E_{det}|^2$ with a certain efficiency η_{qe}. Hence the average number of seemingly detected photons $N_{det} \propto P_{det} \Delta t$ during any given time interval Δt including the average number of electronic dark detections N_n is given by

$$N_{det} = \eta_{qe}\left(N_r + N_{LO} + 2\sqrt{N_{LO}N_r}\cos(\Delta\omega t + \Delta\phi)\right) + N_n \quad . \tag{4.1}$$

4.1. Signal enhancement on the regeneration side

If no LO beam is present, then N_{LO} is zero and N_r could be masked by N_n. With the LO beam, however, an interference term arises, whose amplitude is proportional to the product $\sqrt{N_{\text{LO}} N_r}$. This term can in most cases be made larger than the electronic noise and thus the regenerated signal can be recovered from the electronic noise. However, this works only if the phase difference $\Delta\phi = \phi_r - \phi_{\text{LO}}$ is sufficiently small during Δt. If $\Delta\omega = \omega_r - \omega_{\text{LO}} \neq 0$, the result for N_{det} has to be demodulated to retrieve the regenerated signal.

Due to the interference term, Eq. (4.1) predicts a surprising additional power contribution to the detected signal when the local oscillator is turned on and kept in phase with the regenerated light. In this case the detected power is bigger than the sum of the LO power and the regenerated power. Of course, energy conservation is still valid, as the additional power is drawn from the WISP field. The reason for this effect is that in quantum field theory the coupling between the fields is linear. As a direct consequence, the coupling between field powers cannot be linear and the amount of power drawn from one field in an interaction depends on the amplitude of the second field, which is already present.

For completeness one should note that Eq. (4.1) assumes a vanishing mismatch of the LO mode and signal mode. The interference term decreases with increasing mode mismatch.

In principle, an LO could have been used already in the ALPS I experiment. In this experiment it was assured for alignment reasons, that the beam transmitted through the end mirror of the production resonator and the regenerated beam are essentially collinear. Therefore the transmitted beam could have been used as local oscillator. However, controlling the phase difference $\Delta\phi$ would have been difficult, as the transmitted optics at the center of the magnet were not accessible during the experiment and thermally induced changes of the optical path length of the transmitted light would have caused problems.

4. ALPS II - High-precision optical metrology boosts sensitivity

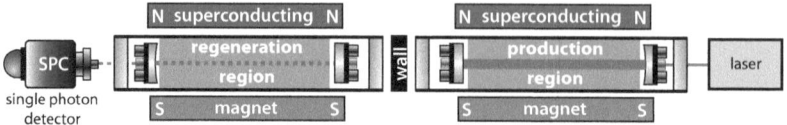

Fig. 4.1: Basic schematic of LSW experiment with regeneration cavity.

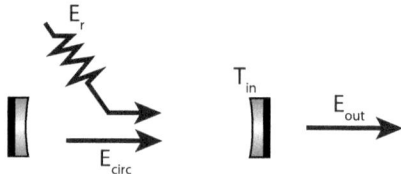

Fig. 4.2: Electric fields in case of a regeneration cavity.

4.1.2. Regeneration cavity

The term *regeneration cavity* denotes another large-scale optical resonator, which comprises the regeneration region of an LSW experiment (see Fig. 4.1). It was first proposed in 1991 by Hoogeveen and Ziegenhagen (while working for the University of Hannover, by the way) [63]. In the following its benefit for an LSW experiment is briefly explained from an optics point of view.

The way how to calculate the transfer function from the incident to the circulating electric field of a general linear optical resonator has been shown by many authors, e.g. [129, 39]. If the cavity mirrors are described by the matrix

$$\begin{pmatrix} \sqrt{1-T_x} & i\sqrt{T_x} \\ i\sqrt{T_x} & \sqrt{1-T_x} \end{pmatrix} ,$$

then the transfer function from the regenerated field to the circulating field in Fig. 4.2 is obtained by exactly the same procedure to be

$$\frac{E_{\text{circ}}}{E_r} = \frac{1}{1 - \sqrt{(1-T_{\text{in}})(1-A_p)} \exp\left(-i\,2\pi\,\frac{\Delta\nu}{FSR}\right)} .$$

4.1. Signal enhancement on the regeneration side

Since the light can only be detected outside of the cavity, it has to leave through one mirror (the other is assumed here to be perfectly reflecting). The electric field outside of the cavity is thus given by

$$\frac{E_{\text{out}}}{E_{\text{r}}} = i\sqrt{T_{\text{in}}}\frac{E_{\text{circ}}}{E_{\text{r}}} \quad . \tag{4.2}$$

If one derives from this equation the transfer function for the optical power by calculating the squared modulus, one arrives at the expression for the power buildup of Eq. (2.15).

Thus the gain of the implementation of the regeneration cavity is its power buildup factor. Furthermore, it was derived already in 2.1.3, that this gain is maximized for the realization of an impedance matched regeneration cavity.

Finally, the technique of signal recycling, which is used in gravitational wave detectors, works very similar to the regeneration cavity.

4.1.3. Possible experimental realizations

Up to now two proposals have been published for the realization of a regeneration cavity enhanced LSW experiment. They are introduced in the following and their sensitivity is roughly compared.

DC detection scheme

This scheme was developed in the scope of this thesis and a detailed explanation is given further down. The detection is done here with a single-photon counter, which integrates all arriving light for a duration of one hour. Regenerated light will thus appear as an excess signal on the integral amount of electronic dark detections.

Heterodyne detection scheme

This scheme was proposed by Mueller et al. [102]. For the realization of a regeneration cavity enhanced LSW experiment two lasers are used here. They emit light with a certain frequency offset of $\Delta\nu \approx 10\,\text{MHz}$, which equals one *FSR* of the regeneration cavity, and which is kept constant by a phase-locked loop. The production cavity resonance is stabilized to the first laser and the regeneration cavity resonance is stabilized to the second. Regenerated light will subsequently appear as a single sideband on the light, which leaks out of the regeneration cavity. It is detected by an ordinary photodiode, whose output signal is demodulated at frequency $\Delta\nu$ to look for a signal. As the correct demodulation phase is not known, one has to demodulate two times with phases differing by $\pi/2$ and add the result quadratically.

Comparison of sensitivities

The demodulation process shifts sideband frequencies of positive and negative demodulation frequency to DC. Thus in the case of the heterodyne detection scheme the regeneration signal, which is a single sideband, is superimposed with the shot noise from both sides of the carrier. Nothing comparable happens in the case of the DC detection scheme, where the regenerated signal is represented by the carrier itself. Thus in the absence of any technical noise source the latter should be more sensitive.

On the other hand it is difficult to find a detector for the DC scheme, whose electronic noise is low enough to be essentially shot noise limited. Maybe a transition edge sensor can achieve that [100]. In contrast, a heterodyne detection scheme can be easily made shot noise limited.

4.2. Basic design study for the ALPS II experiment

In the following basic design considerations for the ALPS II experiment are made to derive approximate values for the most crucial parameters of the optics part.

4.2.1. Cavity design

The regenerated light has to be resonant inside the regeneration cavity in order to be amplified. Thus it has to match the regeneration cavity eigenmode, which can only be achieved efficiently, if each resonator is the extension of the other's eigenmode.

4.2. Basic design study for the ALPS II experiment

Fig. 4.3: Schematic of eigenmodes of production and regeneration cavity for ALPS II.

Accordingly, they should be designed as plano-concave resonators with the plane mirrors at the center. This is depicted in Fig. 4.3.

4.2.2. Power buildup factors

The power buildup of production and regeneration cavity should be as high as possible. In the case of the production resonator an upper limit is given either by the damage threshold of the mirror coatings or by the maximum tolerable thermal effect in the optics. As far as the thermal effects are concerned an upper limit might be derived from the design of Advanced LIGO, where an optical power of 800 kW shall be stored in the arm cavities [50]. On the one hand in case of Advanced LIGO it is planned to compensate thermally induced wavefront distortions with the help of complex compensation systems, which represents efforts that shall not be undertaken for ALPS II. On the other hand, wavefront distortions are of no concern in ALPS II, as here only the resonator's ability to enhance the power level is important. From this discussion a conservative design value for the circulating powers of 150 kW is derived. As the losses of high quality mirror coatings can be $A_m \leq 10$ ppm (and less than half of this should be due to absorption), the mirrors are heated with a power of less than 1 W, which should be acceptable if the beam radius is big enough on their surfaces. With the frequency conversion device presented in Chapter 2 at hand, an incident power level of roughly 30 W should be achievable either for 532 nm or for 1064 nm. The infrared laser source would be the same, which was already used for the ALPS I experiment. These considerations result in a design power buildup for the production cavity of $PB_p = 5000$.

The regeneration cavity will not store significant amounts of power and thus its

4. ALPS II - High-precision optical metrology boosts sensitivity

power buildup is not limited by a damage threshold or thermal effects. Thus its power buildup factor can be larger than PB_p and would be limited by the losses of the mirrors. With the estimated losses given above a power buildup in an impedance matched case of $PB_r = 40000$ should be achievable (see Eq. (2.19)).

4.2.3. Cavity length and free apertures

The power buildup factors just derived can only be obtained if additional losses due to clipping of the beam remain small enough. This is an important issue as far as searches for APs or MCPs are concerned, because the apertures of cryogenic magnets are rather small.

Due to the diffraction of a Gaussian beam this condition translates into a condition for the maximum cavity length and minimum free aperture. As both cavities will have their waist at the inner mirrors, the clipping losses per round trip will be dominated by the outer apertures. In this case the beam, which is clipped least by an aperture of radius r_{ap} and a cavity length d, has a waist radius at the central mirror of

$$w_0 = \sqrt{\frac{\lambda_0 d}{\pi}} \ .$$

The overall fractional losses per round trip are then given by

$$A_p = A_m + \exp\left(-\frac{\pi}{\lambda_0} Q\right) \quad \text{with} \quad Q = \frac{r_{ap}^2}{d} \ . \tag{4.3}$$

These losses are plotted in Fig. 4.4 as a function of the parameter Q for assumed losses per mirror by scattering and absorption of 8 ppm. One can see that the power buildup factors chosen above are possible with such mirrors. Obviously the clipping losses rise very rapidly below a certain threshold value of Q. Due to its smaller divergence angle, 532 nm light would be a clearly better choice here.

Another way to demonstrate these results is shown in Fig. 4.5. Here Eq. (4.3) was solved for r_{ap} under the condition that the chosen values for the power buildup factors are achieved. Obviously the necessary aperture grows rapidly with the cavity length. As the maximum realistic free aperture radii of cryogenic magnets are on the order of 25 mm, the maximum cavity length is limited to about 160 m for 1064 nm

4.2. Basic design study for the ALPS II experiment

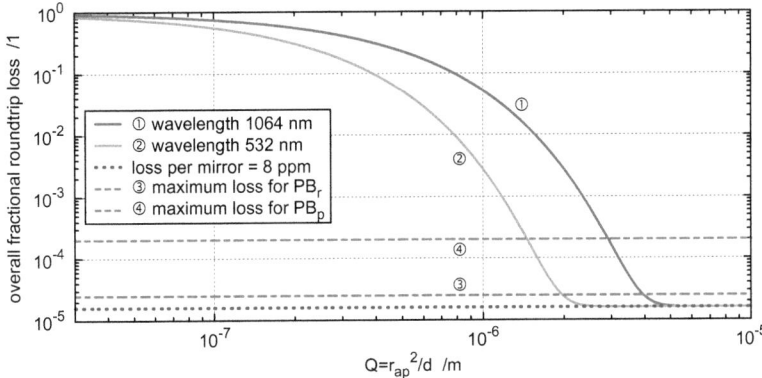

Fig. 4.4: Overall round trip losses due to mirror losses ($A_m = 8$ ppm) and due to clipping for two laser wavelengths. The contribution of the mirrors alone and the maximum losses allowed for achievement of the design values of the power buildup factors are also shown as broken lines.

and to about 320 m for 532 nm. The corresponding linewidth of such a regeneration cavity is on the order of only 10 Hz.

As these lengths are upper limits one should include some safety margin into the calculations for a realistic experiment. Thus cavity lengths of 100 m for infrared and 200 m for visible light are proposed here, both requiring a minimum free aperture of $r_{ap} = 20$ mm. As the sensitivity of an LSW experiment for low mass WISPs is strongly dependent on these lengths, an experiment with visible light would ideally achieve a significantly higher sensitivity than an experiment with infrared light.

4.2.4. Highest intensity

The highest intensity will occur on the central plane mirror of the production resonator. For the proposed cavity lengths just mentioned the waist size on this mirror

4. ALPS II - High-precision optical metrology boosts sensitivity

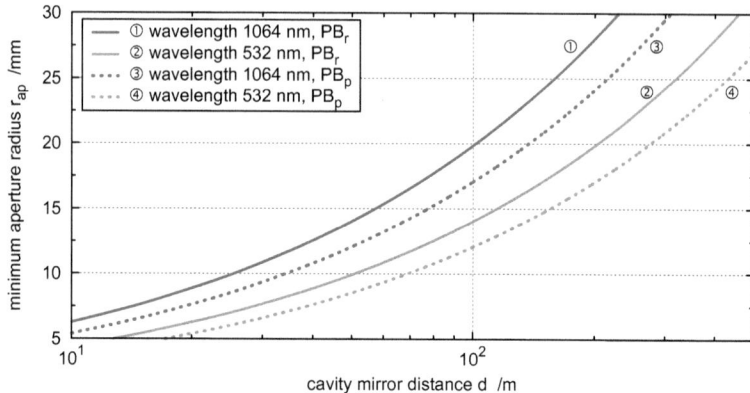

Fig. 4.5: The minimum allowed aperture for the wavelengths and power buildup factors given in the legend as a function of cavity length. The other parameters are the same as in Fig. 4.4.

would be $w_0 = 5.9\,\text{mm}$ for both wavelengths. With the power buildup of the production cavity the peak intensity on this mirror is $270\,\frac{\text{kW}}{\text{cm}^2}$. This value is well below the laser-induced damage thresholds of most dielectric mirror coatings, especially those of very high-quality optics coated with the IBS technique.

4.2.5. Spatial overlap of cavity modes

In order to assure a good matching of the regenerated mode to the eigenmode of the regeneration cavity their relative alignment has to be correct. Because the mode of the regenerated light is the same as the eigenmode of the production cavity, this condition translates into a condition for the overlap of the two eigenmodes. The fraction of the regenerated power, which remains resonant at the TEM_{00} mode of the regeneration cavity is given by η_{00}. With the definition, that the light propagates along the z-axis, η_{00} can be easily estimated for small relative translations along and

4.2. Basic design study for the ALPS II experiment

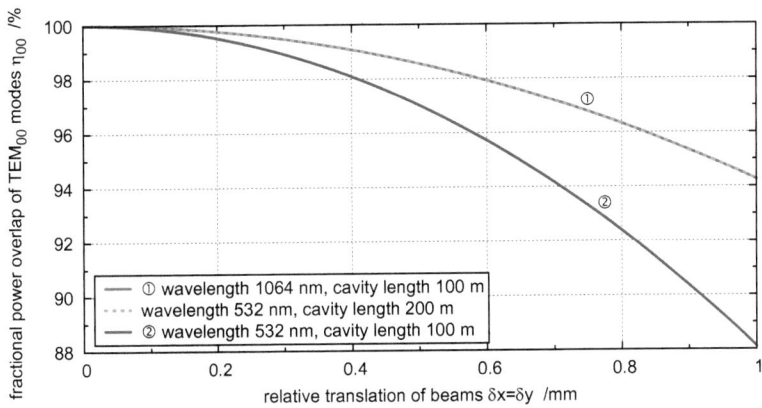

Fig. 4.6: The fractional power overlap of production and regeneration cavity mode as a function of relative lateral translation. To represent a worst-case, the same amount of translation is assumed here for *both* lateral dimensions.

tilts around the x- and y-axis with the help of Eq. (2.9) and Eq. (2.11). Fig. 4.6 shows the result for a relative translation of the cavity eigenmodes of equal amounts in *both* dimensions. When the cavity length is optimized for each wavelength, the mode's waist size w_0 is the same for both wavelengths. Hence the sensitivity to translations is also the same for both wavelengths. Additionally, the case of equal cavity lengths for the two wavelengths is shown. Due to the large beam diameter, a relative translation of the eigenmodes should be of no concern in ALPS II.

Fig. 4.7 shows the dependence of η_{00} on a relative angular tilt. Here a completely different result is obtained as relative angular tilts as tiny as shown in this figure are difficult to achieve. Moreover, selection of a shorter wavelength makes the result even worse. As a reduction in η_{00} up to ten percent might be acceptable, angular tilts have to be smaller than 6.3 μrad in the case of a wavelength of 532 nm and smaller than 13 μrad in the case of a wavelength of 1064 nm. It is currently unclear how to achieve such small tilt angles in ALPS II. Maybe bonding and interferometric

183

4. ALPS II - High-precision optical metrology boosts sensitivity

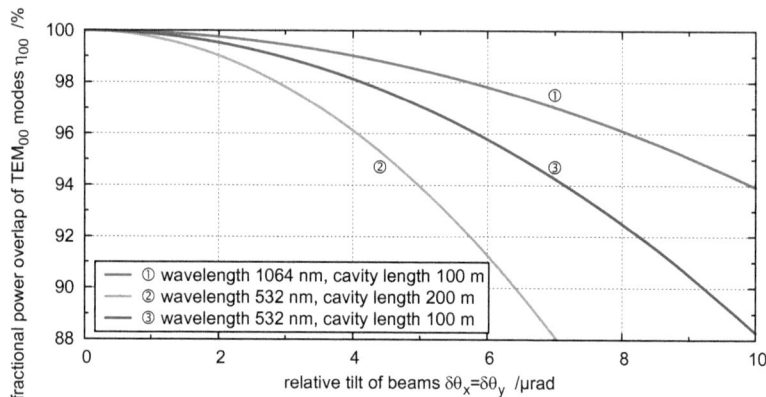

Fig. 4.7: The fractional power overlap of production and regeneration cavity mode as a function of relative tilt. To represent a worst-case, the same amount of tilt is assumed here around *both* lateral axes.

alignment techniques as applied in the construction of the LISA optical bench are an option [135].

As was explained in 4.2.1, the production and regeneration cavity will be designed as plane concave mirrors with their central mirrors being plane. To facilitate the stabilization of the regeneration cavity to the production light, and to enable tests of their relative alignment in situ, additional optical components have to be placed between these plane cavity mirrors. Hence the cavity modes are no longer exact extensions of each other. Correspondingly, the matching of the regenerated light to the eigenmode of the regeneration cavity will be reduced. The same is true for a difference in waist size of the two eigenmodes due to tolerances in the manufacturing of the curved mirrors. However, a similar calculation like the ones above shows that waist distances on the order of 2 m and waist size variations on the order of a few percent are negligible for the power overlap.

4.2.6. Mirror lifetime

To achieve the high power buildups proposed above, the coatings of the cavity mirrors have to have very high quality. Such a high quality is offered by coatings produced via the IBS technique.

In the field of gravitational wave detection lots of experience has been acquired on the long-term stability and laser-induced damage thresholds of IBS coatings at a laser wavelength of 1064 nm. From this experience it is expected that IBS coatings for this wavelength will show sufficiently long life times at the power levels proposed above.

Considerably less experience has been acquired with high-power applications of such coatings for a wavelength of 532 nm. Additionally, for lower quality coatings a rather short life time was observed in the ALPS I experiment (see 3.5.2). Therefore a simple long-term test was conducted with an identical experimental setup as it is described in 2.4.4 and shown in Fig. 2.21. Here, the original mode analyzer cavity was exchanged by an essentially identical device, which was fabricated in a low class clean room to reduce contamination of its coatings as far as possible. Its interior was sealed completely when the mirrors were glued to the spacer. By this means, it was assured that no dust or gas from the outside could contaminate the mirror coatings after leaving the clean room. The same glue was used since years for the production of similar resonators for a wavelength of 1064 nm and caused no major problems.

This resonator was equipped with a frequency stabilization control loop and automatic lock acquisition electronics. Thus the harmonic power at 532 nm could be made resonant inside the cavity over long timescales. The incident harmonic power and the power transmitted through the curved cavity mirror were constantly monitored to determine the power buildup. The result is presented in Fig. 4.8.

Obviously, the mirror coatings of this resonator had a limited life time of approximately 250 h. The apparent rise of the power buildup in the beginning was in fact caused by a rise of the detected transmitted power. The corresponding photodetector was positioned right behind the resonator mirror and this mirror showed

4. ALPS II - High-precision optical metrology boosts sensitivity

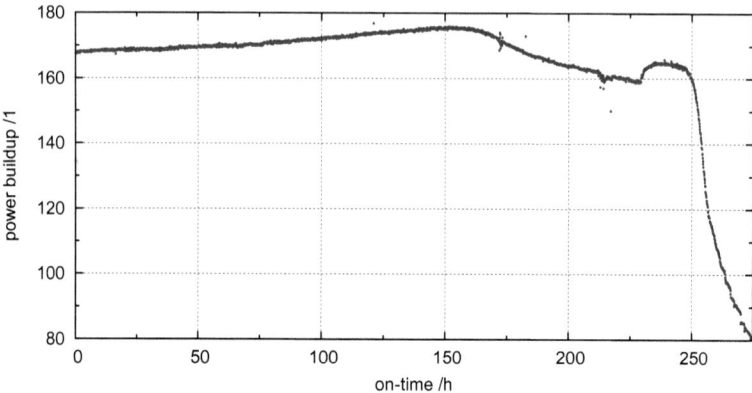

Fig. 4.8: Result of a long-term power buildup measurement at a wavelength of 532 nm. The circulating power was about 400 W, the highest intensity on a mirror surface was 600 $\frac{kW}{cm^2}$. The resonator was nearly impededance matched.

a dramatically increased scattering of green light after the power buildup had degraded. Thus it is assumed here, that the rise in this measurement was caused by a steadily increasing amount of scattered light from this mirror surface over time, and that the same process caused the strong power buildup degradation in the end. The deeper nature of this process is not known to the author.

Of course, this result is not a general proof of a limited life time of the applied coatings at a wavelength of 532 nm. But it represents at least a hint that such problems might exist. Therefore it was decided to use 1064 nm light in the ALPS II experiment.

For further testing a similar resonator should be prepared, which avoids the use of glue or a piezo, as these components are suspects for contamination of the mirror surfaces.

4.3. A proposed experimental setup

In order to benefit from the implementation of a regeneration cavity, its resonance has to be stabilized to the frequency of the regenerated light. If one uses a local oscillator, it can be used for the stabilization. However, if no local oscillator beam is used, the setup has to assure, that the power of any not regenerated light, which arrives at the detector is much smaller than the detector's noise equivalent power P_{ne}. One should note here, that one expects

$$P_{\text{ne}} \leq 3.4 \times 10^{-21}\,\text{W} \quad ,$$

as the single photon detector of the ALPS I experiment showed a dark flux of only $32.4\,\frac{\text{photons}}{\text{h}}$!

4.3.1. Proposed experimental setup for ALPS II

The experimental setup, which was proposed for the ALPS II experiment, and which was to a large extend developed in the context of this thesis is presented in Fig. 4.9.

The crucial idea is the utilization of two different wavelengths. The single-photon detector is protected by dispersive and dichroic optics, such that it can only be hit by the regenerated wavelength. The regeneration cavity mirrors CBS2 and REM are coated such that the cavity has a very high power buildup for the regenerated wavelength and a low one for the other. Hence the other wavelength can be used to stabilize the cavity. For simplicity this stabilization wavelength is again obtained via second harmonic generation from the infrared light.

General structure

As regeneration and production resonator are planned to be very long the overall optical setup is distributed onto three optical tables. On optical table 1 the injection stage for the infrared light is located together with the input coupler of the production cavity PIC.

4. ALPS II - High-precision optical metrology boosts sensitivity

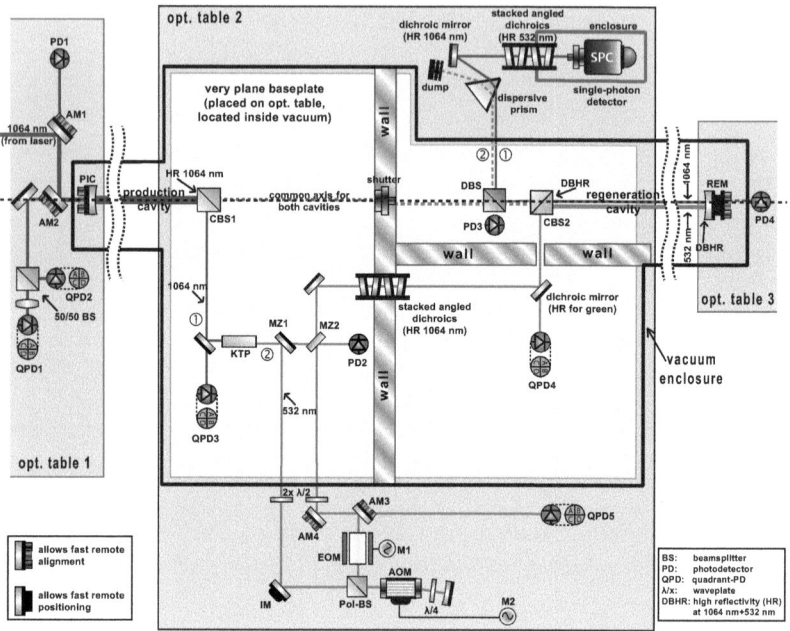

Fig. 4.9: Rough schematic of the proposed experimental setup of production and regeneration cavity for the ALPS II experiment. Red lines (1) mark light with wavelength 1064 nm and green lines (2) mark light with wavelength 532 nm.

The second table encompasses a rigid very plane central board (e.g. made from Zerodur or ALPLAN), on which the inner cavity mirrors CBS1 and CBS2 are mounted. These mirrors are realized as very plane-parallel, coated beamsplitter cubes, such that they do not divert the transmitted light from the WISP path, which is the common axis of both cavities (the same has to hold for DBS). The board is divided into a left and a right half by walls, to separate the infrared light of the production cavity

4.3. A proposed experimental setup

from the regenerated light. The converted light can pass this wall through a set of dichroic optics, which does not transmit any significant amount of infrared light. A shutter (or several successive ones) in this wall can be opened to test the mode overlap between production and regeneration cavity in situ with the help of PD3 and PD4. To accomplish that the plane-parallel cube DBS is designed to transmit a few percent of its incident light.

Some auxiliary optics, the single-photon counter and its protective dichroic and dispersive optics are also placed on the second table. On the third table the far-end mirror of the regeneration cavity and one detector for the in situ mode overlap test are located.

The central board and both long cavities are contained in a single vacuum enclosure to protect the optics from contamination and to avoid noise from air currents and acoustics.

Wavelength selection and frequency control scheme

The stabilization of the frequency of the incident light to the resonance of the production cavity is done in the same way as in ALPS I. The laser frequency is the actuator of the control loop. Its error signal is derived from a demodulation of PD1.

A small fraction of the infrared light, which is circulating inside the production resonator is transmitted through the inner cavity mirror CBS1. A fraction of this transmitted light is converted to 532 nm by a KTP crystal. This crystal is mounted such that its optical z axis is oriented under an angle of 45 deg to the optical table. By this, conversion from vertical and horizontal infrared polarization is possible, which is necessary to search for certain WISP species. The main part of the green light then leaves the vacuum and the central board.

Outside of the vacuum some optics are located, which are either hardly vacuum compatible or have to be accessible by the experimenter. First the beam's polarization is corrected by a pair of $\lambda/2$ waveplates depending on the infrared polarization, and the beam may also be attenuated in combination with Pol-BS. An AOM in double-pass allows for a fixed frequency shift M2 and an EOM imprints phase mod-

4. ALPS II - High-precision optical metrology boosts sensitivity

ulation sidebands at frequency M1 for the stabilization of the regeneration cavity.

Then the green light is send back into the vacuum and its outside acquired phase noise is read out via a Mach-Zehnder interferometer formed by a highly reflecting mirror MZ1, a weakly reflecting plate MZ2 and by detector PD2. After demodulation of the signal from PD2 at two times the frequency M2, the piezo driven mirror IM is used to compensate for this phase noise. After that the green beam leaves the production side of the central board.

On the regeneration side the green light is injected into the regeneration cavity through its inner cavity mirror CBS2. The length of the regeneration cavity is stabilized to the incident light's frequency via its piezo-driven end mirror REM. The error signal for this control loop is obtained by demodulating the integral output of quadrant PD QPD5 with frequency M1. Any green light, which leaves the regeneration cavity into the direction of the single-photon detector is split off the regenerated light by the dispersive and dichroic optics, which protect the single-photon detector.

The two employed wavelengths differ exactly by a factor of two. Furthermore, the phase stable SHG process, the Mach-Zehnder interferometer, the vacuum and the rigid central board assure sufficient phase stability between infrared and green light. Hence for ideal cavity mirrors every second resonance of the harmonic wavelength will also be a resonance of the fundamental one.

Finally, as was found in [82], the mirrors are not ideal. The penetration depths of the two wavelengths circulating inside the regeneration cavity into the mirror coatings differ by an amount, which is stationary but varies from mirror to mirror. Thus in general the resonance frequencies will not be the same. Their constant frequency offset has to be measured once with open shutter, and frequency M2 has to be adjusted to compensate for this.

Alignment control scheme

As the three optical tables will move relative to each other, control loops have to be set up, which continuously reestablish optimum alignment of the whole setup. Only minor parts of this control scheme were developed in the scope of this thesis and it is therefore only briefly summarized here.

The spatial reference is defined by the positions and angular alignment of CBS1, CBS2, QPD3 and QPD4 which are fixed to the central board. These define the axis of each cavity and they will be aligned initially such that the two axes are collinear. The central board is assumed to be stable enough to give a sufficiently precise reference.

Once the eigenmodes of the cavities are established, the technique of differential wavefront sensing will be used via QPD1-5 to maintain the eigenmodes themselves and the alignment between incident beams and eigenmodes.

4.3.2. Technical challenges

Examples for technical challenges of this design, which need to be investigated further in the future are the following:

- **Plane parallelism of optics:** A wedge angle of the optics CBS1, CBS2 or DBS would result in a tilt of the transmitted beam relative to the WISP beam. Such a tilt would be a problem, as the transmitted beam will be used to characterize and maybe even to align the overlap between production and regeneration cavity. From Eq. (3.5) it can be seen that for usual optics the tilt angle becomes approximately half the wedge angle, which has to be compared with the tilt sensitivity of the long cavities, shown in Fig. 4.7. Hence one would need optics with wedge angles significantly smaller than 10 µrad or 2 arcsec to keep their influence negligible. As so-called etalon substrates are commercially available with wedge angles below 1 arcsec, this problem appears to be solvable.

- **Down-conversion:** The discrimination between the two wavelengths applied in this setup works only if the green light does not exhibit down-conversion into infrared light. Even a tiny effect can be a problem here due to the especially tiny value of the noise equivalent power P_{ne} of the single-photon detector, which was given at the beginning of this section. This might be caused for instance by

4. ALPS II - High-precision optical metrology boosts sensitivity

fluorescence after optical absorption. Some experiments, which were not done by the author, could already measure such effects in usual dispersive optics or absorbing spectral filters. An exact quantification and spectral analysis is not know to the author and has to be done in the future.

4.4. Projected sensitivity

In this section the sensitivity of the ALPS II experiment is estimated as it was proposed in this chapter. The estimation was done exemplarily for the coupling constants $g_{-/+}$, which describe the coupling of axion-like particles (APs) to photons. The scaling of these constants as a function of the parameters discussed in this chapter is for low AP masses given by (see Eq. (3.1) and [102])

$$g_{-/+} \propto \frac{1}{BL_m} \sqrt{\frac{1}{\eta_{ovl}}} \sqrt[4]{\frac{1}{\lambda_0 \, PB_p \, PB_r \, P_{inc}}} \quad . \tag{4.4}$$

Here B is the magnetic field strength, L_m the length of the magnetic field region on each side of the wall and η_{ovl} denotes the spatial overlap of production and regeneration cavity mode. The coupling constants for the other WISP types of interest scale in a similar manner. Tbl. 4.1 lists the values, which were assumed for the parameters in Eq. (3.1).

Fig. 4.10 shows the projected sensitivity. It compares the published results for the ALPS I experiment with the predicted results for two versions of a future ALPS II experiment. The first version assumes utilization of the same detector and data taking duration as for the ALPS I results. Obviously incorporation of the optical metrology techniques described in this chapter boosts the sensitivity of the LSW experiment by orders of magnitude. It drops even below the results of the CAST experiment. This means it explores parameter space, which no experiment has probed before. For the case of MHPs (which is not shown here) this applies even for the whole parameter space from the ALPS I results down to the projected ALPS II results.

For the second projection of the ALPS II sensitivity it was assumed that a shot noise limited detector was available and that the data taking duration was increased

parameter	value
B	5.0 T
L_m	100.0 m
η_{ovl}	0.9
λ_0	1064 nm
P_{inc}	30.0 W
PB_p	5000
PB_r	40000

Table 4.1: Assumed parameters of a future ALPS II experiment.

to 50 h for this single WISP species. Such a shot noise limited detector might possibly be realized by utilization of a so-called transition edge sensor [100]. In this case the sensitivity gets close to a region in parameter space, whose corresponding particles could explain a surprisingly high transparency of the universe for very high energy radiation observed by the MAGIC collaboration (see 3.2.2).

4. ALPS II - High-precision optical metrology boosts sensitivity

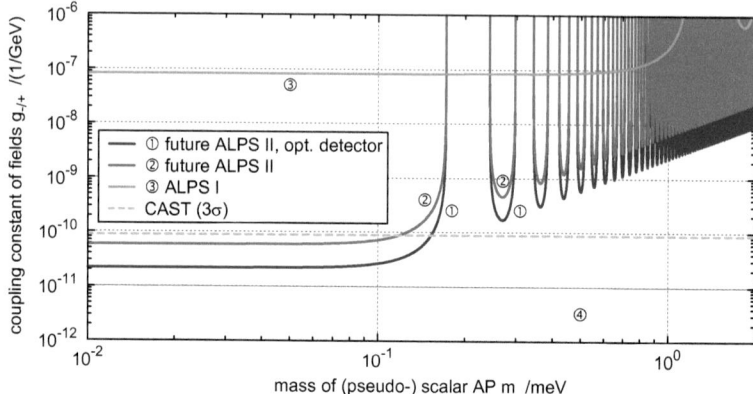

Fig. 4.10: Projected sensitivities (95 % confidence intervals) of two versions of the future ALPS II experiment compared to the current ALPS I sensitivity. Additionally, the sensitivity of the CAST experiment is shown (only as guide for the eye because it is a 99.7 % confidence interval result). The yellow region (4) marks those APs, which could explain the surprisingly high transparency of the universe for very high energy radiation (see 3.2.2).

4.5. Summary and outlook

The ALPS collaboration decided to plan and build a new, enormously improved LSW experiment, which shall incorporate a regeneration cavity. It was dubbed ALPS II.

In this chapter first some techniques were explained, which have the ability to enhance the sensitivity of an LSW experiment (Section 4.1). The most important one is the regeneration cavity. Two different ways to realize such a cavity were briefly compared.

A basic design study was made to find suitable values for the most important parameters of the optical part of the ALPS II experiment (Section 4.2). A length of

4.5. Summary and outlook

production and regeneration resonator of 100 m each appeared feasible. With laser light at 532 nm these could be made even twice as long. Strong constraints were derived on relative tilts between production and regeneration cavity mode. These are on the order of 10 µrad. A simple long-term test of the life time of mirror coatings at 532 nm was conducted and yielded disappointing results. Hence it was proposed to design the experiment for 1064 nm.

A basic experimental setup of the optics part of the ALPS II experiment was presented (Section 4.3). It utilizes a two wavelength scheme to stabilize the regeneration cavity to the regenerated light, without influencing the single-photon detector. Some technical difficulties like possible down-conversion were briefly described.

Finally, the projected sensitivity of such a setup was presented and compared to other experiments (Section 4.4). Its sensitivity was boosted by orders of magnitude by the incorporation of several techniques from optical metrology.

If the ALPS II experiment is built, it would probe big regions of parameter space, which have never been probed before. Therefore it has good chances to detect WISPs and thus revolutionize our understanding of the universe.

Chapter 5

Conclusion

The availability of high-power 532 nm laser radiation offers the chance to enhance the efficiency of engineering applications as well as the sensitivities of many physics experiments, which were previously conducted in the infrared. Examples for such experiments are gravitational wave detectors or light shining through a wall (LSW) experiments. In the latter case, the combination of such a laser with optical resonators achieves a break-through in sensitivity.

In this thesis, first, a technique was developed, which allows to reduce the intensities inside the nonlinear crystal of an external-cavity second harmonic generation (SHG) scheme without sacrificing its external conversion efficiency (Chapter 2). This technique facilitates the reduction of intensity dependent absorption effects, which in the past were a limiting factor for the harmonic power levels achieved from many nonlinear materials. This was tested for a setup utilizing a PPKTP crystal and resulted in a constant power of 5.3 W, which is the highest long-term stable single-frequency 532 nm power from this material reported so far. A model was developed, which adequately linked the remaining deviations of the crystal performance from the ideal case to thermal dephasing processes caused by linear absorption and gray tracking.

The experiences with the PPKTP experiment set the basis for the development

5. Conclusion

of a 134 W CW 532 nm metrology laser source. The extraordinary high power was emitted at a single frequency and at least 97 % of it was contained in the TEM_{00} mode. The nonlinear material applied here was LBO. This SHG device represents a break-through in the amount of CW single-frequency power available at this wavelength. In fact it is the by far highest level reported to date. The external conversion efficiency of 90 % was also the best reported so far for high-power devices. Despite the high power levels involved, the output power was limited only by the available amount of infrared light. Hence there is a good chance to generate considerably more harmonic power with this SHG scheme in the future, when more powerful infrared lasers become available.

To derive the fundamental mode content of the harmonic power with high precision, a new measurement technique was developed in this thesis, which helps to reduce the uncertainty of this value compared to other techniques.

Simulation of the LBO device obtained, that this specific setup should be able to achieve external conversion efficiencies of more than 80 % for any incident fundamental power level above only 12 W. It is thus a very interesting frequency conversion device also for any intermediate power infrared laser.

The demonstration of a high power, low noise 532 nm laser source is important as it opens up a new domain of design parameters for future experiments, as for example third generation gravitational wave detectors.

The detection of gravitational waves would open up a fundamentally new way to observe astrophysical processes in our universe.

Benefits for one field of research obtained from important progress in another are common in science and often have led to break-through developments. A recent example of such a fruitful combination of generally dissimilar fields of science was also presented in this thesis. Here the joining of precision optical metrology and hypothetical particle physics led to a huge step forward in the sensitivity of LSW experiments. They search for so-called weakly interacting slim particles (WISPs), which are not part of the Standard Model of particle physics.

Such experiments traditionally used pulsed lasers as sources for their primary particles, namely photons. In this thesis the first LSW experiments were presented, where the pulsed laser was substituted by a frequency doubled continuous-wave source, which was stabilized to a large-scale optical resonator (Chapter 3).

First, a proof of principle experiment was set up with the additional goal to gain important design parameters for an optimized LSW experiment. The design, implementation, characterization and maintenance of the optical injection stage, the harmonic generator and of the 9 m long production resonator of this initial phase of the ALPS I experiment were described.

Based on the presented findings, in the next step the ALPS I experiment was upgraded to maximize its sensitivity. This upgrade process paved the way to the most stringent exclusion limits world-wide from laboratory experiments for a certain class of WISPs. The implementation of a resonant high-power SHG stage and of a narrow-linewidth production resonator was described in the scope of this thesis. These were the crucial steps to improve the upper limits by a factor of two to four in broad regions of the parameter space. A very high optical power of 1 kW at a wavelength of 532 nm was stored on average inside the optical resonator for an integral data taking duration of 55 h. Finally, these numbers demonstrate beyond any doubt the superiority of a resonator-enhanced LSW experiment above the traditional pulsed laser based schemes.

On the one hand evidence for the existence of a certain class of WISPs is cumulating and therefore a finding might be just around the corner. On the other hand there are also several models, which predict WISPs with coupling constants to photons, which are still way off the sensitivity of current LSW experiments. The implementation of a second large-scale resonator denoted as regeneration cavity, and the use of a local oscillator, allow to considerably increase the sensitivity of an LSW experiment. Hence it was decided to plan and build the ALPS II experiment with resonator lengths of several tens of meters each.

In this thesis the dependence of the performance of ALPS II on several important experimental parameters was investigated (Chapter 4). This showed again, that the application of 532 nm light would theoretically lead to a higher sensitivity than

5. Conclusion

utilization of infrared light. However, a simple experiment demonstrated that the life-times of mirror coatings for 532 nm laser light might be limited. An unprecedented basic design for the control of the regeneration cavity was developed, which ideally does not influence the employed single-photon detector.

The sensitivity of a realistic ALPS II experiment was estimated here. It turned out to be orders of magnitude higher than that of all LSW experiments, which have been implemented to date. Thus its realization has good chances to detect WISPs. Such an event would start a new era in particle physics and would fundamentally change our understanding of the universe.

Appendix A

Basic optics

A.1. Maxwell's equations

It is fascinating that the propagation of any electro-magnetic wave (i.e. any light wave) through an arbitrary medium is completely described by a single set of differential equations, namely Maxwell's Equations [14, 105, 40]

$$\nabla \mathbf{B}(\mathbf{r},t) = 0, \qquad \nabla \mathbf{D}(\mathbf{r},t) = \rho_c(\mathbf{r},t), \qquad (A.1)$$
$$\nabla \times \mathbf{E}(\mathbf{r},t) + \frac{\partial}{\partial t}\mathbf{B}(\mathbf{r},t) = 0, \qquad \nabla \times \mathbf{H}(\mathbf{r},t) - \frac{\partial}{\partial t}\mathbf{D}(\mathbf{r},t) = \mathbf{J}_c(\mathbf{r},t),$$

with

$$\mathbf{B}(\mathbf{r},t) = \mu_0\left(\mathbf{H}(\mathbf{r},t) + \mathbf{M}(\mathbf{r},t)\right), \qquad \mathbf{D}(\mathbf{r},t) = \epsilon_0 \mathbf{E}(\mathbf{r},t) + \tilde{\mathbf{P}}(\mathbf{r},t), \qquad (A.2)$$
$$\nabla \mathbf{J}_c(\mathbf{r},t) = -\frac{\partial}{\partial t}\rho_c(\mathbf{r},t), \qquad \mathbf{E}(\mathbf{r},t) = \underline{\sigma}\mathbf{J}_c(\mathbf{r},t),$$

where $\mathbf{E}$, $\mathbf{B}$, $\mathbf{D}$, $\tilde{\mathbf{P}}$, $\mathbf{H}$ and $\mathbf{M}$ denote the electric, magnetic, displacement, atomic polarization, magnetizing and magnetization field respectively, and ρ_c and $\mathbf{J}_c$ are the charge and current density. Here SI units were used and the electric and magnetic constants are defined by the vacuum speed of light as $c_0 = 1/\sqrt{\mu_0 \epsilon_0}$ and $\underline{\sigma}$ is meant to be a tensor. Within the scope of this thesis above equations can be simplified

A. Basic optics

considerably. Typical transparent media are isolators without free charges

$$\rho_c = 0 \quad , \quad \mathbf{J}_c = \mathbf{0} \quad ,$$

they are not ferromagnetic (introducing the linear magnetic susceptibility, which is assumed here as frequency independent for simplicity)

$$\mathbf{M} = \chi_m^{(1)} \mathbf{H} \quad ,$$

and it is sufficient to deal with only one dimension of propagation

$$\mathbf{r} \to z \quad .$$

With these simplifications the propagation of the electric field component of a light wave is described by

$$\nabla^2 E(z,t) - \frac{1+\chi_m^{(1)}}{c_0^2}\frac{\partial^2}{\partial t^2}E(z,t) = \frac{1+\chi_m^{(1)}}{\epsilon_0 c_0^2}\frac{\partial^2}{\partial t^2}\tilde{P}(z,t) \quad , \tag{A.3}$$

and an analogous equation describes the propagation of the magnetic field component. Because one finds that the amplitudes of the two components compare like

$$|E| = c_0 |B| \quad ,$$

the contribution of the magnetic field is negligible to the effects analyzed in this thesis and thus it is not investigated further (nonetheless the energy density of the electromagnetic field is distributed equally over the electric and magnetic component).

A.2. General solution describing light propagation

A general solution to the above differential Eq. (A.3) in a linear medium or vacuum with

$$\tilde{P}(t) = \epsilon_0 \chi_e^{(1)}(\nu) E(t) \quad ,$$

where $\chi_e^{(1)}(\nu)$ denotes the linear electric susceptibility is given by

$$E(z,t) = \frac{1}{2} U(x,y,z,t)\, e^{i(kz - 2\pi\nu t)} + (cc) \quad , \tag{A.4}$$

which describes a monochromatic electromagnetic wave, and which is necessarily real because the electric field is a measurable quantity. The quantity U in Eq. (A.4) is called slowly varying complex amplitude because it encompasses only variations of the wave's amplitude and phase that are much slower than ν. The parameter k is called the wave vector and given by $2\pi \mathrm{n}(\nu)/\lambda_0$ with λ_0 denoting the laser wavelength in vacuum and $\mathrm{n}(\nu)$ the refractive index of the medium.

Following the superposition principle for linear operators the right complex conjugate part (often abbreviated by '*(cc)*') can be simply ignored whenever equations are considered, which make use of *linear* operators only. In such cases all calculations can be performed with the remaining complex exponential function (which is much easier than involving trigonometric algebra) and the transition to a real expression can be postponed to the end and done just by adding the complex conjugate again [105]. In order not to loose factors of 2 it is important to obey this convention (and *not* to obtain the complex field first by dropping the conjugate part and later return to a real field just by dropping the imaginary part again). At the same time the correct definition of the field's intensity (see below) has to be used. Also note, that this thesis deals with nonlinear effects, which means that the complex conjugate part is essential at some points, where *nonlinear* operators are used (like the square operator applied to the electric field)!

A.3. Intensity and power

The amount of energy per unit time, which is carried by a light wave described by Eq. (A.4) is usually expressed in terms of intensity I and power P. The calculation of these quantities is a typical example of a problem involving nonlinear operators. Thus application of the convention for such cases defined below Eq. (A.4) leads to the following definitions for the electric field under investigation in its complex and

A. Basic optics

real versions and for the corresponding expressions for its intensity [105, 39]:

$$\text{Real el. field:} \quad \mathbf{E}(x,y,z,t) = \frac{1}{2}\mathbf{U}(x,y,z,t)\,e^{i(\mathbf{kr}-2\pi\nu t)} + (cc) \quad \text{(A.5)}$$

$$= \text{Re}\left\{\mathbf{U}(x,y,z,t)\,e^{i(\mathbf{kr}-2\pi\nu t)}\right\} \quad,$$

$$\text{Intensity:} \quad I(x,y,z) = |\langle \mathbf{S}\rangle_t| = \left|\left\langle \mathbf{E}\times\mathbf{H}^\dagger\right\rangle_t\right|$$

$$= \frac{1}{2}\left|\sqrt{\frac{\epsilon_r\epsilon_0}{\mu_r\mu_0}}\,|\mathbf{U}|^2\,\frac{\mathbf{k}}{|\mathbf{k}|}\right| \approx \frac{1}{2}n\epsilon_0 c_0\,|\mathbf{U}|^2 \quad.$$

Here **S** denotes the Poynting vector. The notation $\langle x \rangle_t$ means averaging of x in time over a period of the oscillation. Finally, in the last equation the relative permeability μ_r was assumed to be very close to unity, which is true for all media and effects relevant in the context of this thesis.

The optical power of the beam is then obtained from the intensity by integration over the complete transversal plane

$$P = \iint\limits_{-\infty}^{\infty} I(x,y)\,\mathrm{d}x\,\mathrm{d}y = \int\limits_0^\infty r \int\limits_0^{2\pi} I(r,\phi)\,\mathrm{d}\phi\,\mathrm{d}r \quad. \quad \text{(A.6)}$$

Bibliography

[1] S. A. Abel, J. Jaeckel, V. V. Khoze, and A. Ringwald. Illuminating the hidden sector of string theory by shining light through a magnetic field. *Physics Letters B*, 666:66–70, 2008.

[2] A. Abramovici and J. Chapsky. *Feedback control systems: a fast track guide for scientists and engineers*. Kluwer Academic Publishers, 2000.

[3] S. L. Adler. Photon Splitting and Photon Dispersion in a Strong Magnetic Field. *Annals of Physics*, 67:599–647, 1971.

[4] S. L. Adler, J. Gamboa, F. Méndez, and J. López-Sarrión. Axions and 'light shining through a wall': A detailed theoretical analysis. *Annals of Physics*, 323:2851–2872, 2008.

[5] A. Afanasev, O. K. Baker, K. B. Beard, G. Biallas, and et al. New Experimental limit on Optical Photon Coupling to Neutral, Scalar Bosons. *Physical Review Letters*, 101:120401, 2008.

[6] M. Ahlers, H. Gies, J. Jaeckel, J. Redondo, and A. Ringwald. Light from the hidden sector: Experimental signatures of paraphotons. *Physical Review D*, 76:115005, 2007.

Bibliography

[7] M. Ahlers, H. Gies, J. Jaeckel, J. Redondo, and A. Ringwald. Laser experiments explore the hidden sector. *Physical Review D*, 77:095001, 2008.

[8] J. Albert, E. Aliu, H. Anderhub, and et al. Very-High-Energy Gamma Rays from a Distant Quasar: How Transparent Is the Universe? *Science*, 320:1752, 2008.

[9] C. Amsler and et al. Review of Particle Physics. *Physics Letters B*, 667:1, 2008.

[10] S. J. Asztalos, G. Carosi, C. Hagmann, D. Kinion, K. van Bibber, M. Hotz, L. J Rosenberg, G. Rybka, J. Hoskins, J. Hwang, P. Sikivie, D. B. Tanner, R. Bradley, and J. Clarke. Squid-based microwave cavity search for dark-matter axions. *Phys. Rev. Lett.*, 104(4):041301, Jan 2010.

[11] S. Avino, E. Calloni, A. Tierno, B. Agrawal, R. De Rosa, L. Di Fiore, L. Milano, and S. R. Restaino. Low-noise adaptive optics for gravitational wave interferometers. *Classical and Quantum Gravity*, 23:5919–5925, 2006.

[12] T. Baer. Large-amplitude fluctuations due to longitudinal mode coupling in diode-pumped intracavity-doubled Nd:YAG lasers. *Journal of the Optical Society of America B*, 3(9):1175, 1986.

[13] R. Ballou and et al. Summary of OSQAR First Achievements and Main Requests for 2008. *CERN-SPSC-2007-039S; PSC-M-762*, 2007.

[14] Y. B. Band. *Light and Matter - Electromagnetism, Optics, Spectroscopy and Lasers (corrected reprint)*. Wiley, 2007.

[15] M. Bass, P. A. Franken, J. F. Ward, and G. Weinreich. Optical Rectification. *Physical Review Letters*, 9(11):446, 1962.

[16] R. Battesti, B. Pinto Da Souza, S. Batut, and et al. The BMV experiment: a novel apparatus to study the propagation of light in a transverse magnetic field. *European Physical Journal D*, 46:323–333, 2008.

[17] H. J. Baving, H. Muuss, and W. Skolaut. CW Dye Laser Operation at 200 W Pump Power. *Applied Physics B*, 29:19–21, 1982.

[18] C.L. Bennett, M. Bay, M. Halpern, G. Hinshaw, and et al. The Microwave Anisotropy Probe (MAP) Mission. *The Astrophysical Journal*, 583:1, 2003.

[19] L. Bergström and A. Goobar. *Cosmology and Particle Astrophysics (2nd edition)*. Springer-Verlag, 2006.

[20] G. Bertone, D. Hooper, and J. Silk. Particle dark matter: Evidence, candidates and constraints. *Physics Reports*, 405:279–390, 2005.

[21] I. I. Bigi. *CP Violation*. Cambridge University Press, 2001.

[22] E. D. Black. An introduction to Pound-Drever-Hall laser frequency stabilization. *American Journal of Physics*, 69:79, 2001.

[23] D. G. Blair. *The detection of gravitational waves (1st edition)*. Cambridge University Press, 1991.

[24] C. Bogan and P. Kwee. internal report, albert-einstein-institute hannover, germany, 2011.

[25] G. D. Boyd and D. A. Kleinman. Parametric Interaction of Focused Gaussian Light Beams. *Journal of Applied Physics*, 39:3597, 1968.

[26] R. W. Boyd. *Nonlinear optics (2nd edition)*. Academic Press (Elsevier), 2003.

[27] G. C. Branco, L. Lavoura, and J. P. Silva. *CP Violation*. Oxford University Press, 1999.

[28] C. Burrage, J. Jaeckel, J. Redondo, and A. Ringwald. Late time CMB anisotropies constrain mini-chraged particles. *Journal of Cosmology and Astroparticle Physics*, 0911:002, 2009.

Bibliography

[29] R. Cameron, G. Cantatore, A. C. Melissinos, G. Ruoso, Y. Semertzidis, H. J. Halama, D. M. Lazarus, A. G. Prodell, F. Nezrick, C. Rizzo, and E. Zavattini. Search for nearly massless, weakly coupled particles by optical techniques. *Physical Review D*, 47(9):3707, 1993.

[30] L. Carrion and J.-P. Girardeau-Montaut. Gray-track damage in potassium titanyl phosphate under a picosecond regime at 532 nm. *Applied Physics Letters*, 77(8):1074, 2000.

[31] CERN. The Large Hadron Collider (LHC), facts and figures. http://public.web.cern.ch, 2008.

[32] Y. Chikashige, R. N. Mohapatra, and R. D. Peccei. Spontaneously Broken Lepton Number and Cosmological Constraints on the Neutrino Mass Spectrum. *Physical Review Letters*, 45(24):1926, 1980.

[33] A. S. Chou, W. Wester, A. Baumbaugh, H. R. Gustafson, and et al. Search for axion-like particles using a variable baseline photon regeneration technique. *Physical Review Letters*, 100:080402, 2008.

[34] MABRILAS collaboration. CUBRILAS: Green Laser Welding - First Results. http://www.ot-mabrilas.de/en/2010/10/20/laserschweisen-mit-grun-erste-ergebnisse, 2010.

[35] J. W. Czarske, R. Phillips, and I. Freitag. Spectral properties of diode-pumped non-planar monolithic Nd:YAG ring lasers. *Applied Physics B*, 61:243–248, 1995.

[36] C. Czeranowsky, E. Heumann, and G. Huber. All-solid-state continuous-wave frequency-doubled Nd:YAG-BiBO laser with 2.8-W output power at 473 nm. *Optics Letters*, 28(6):432, 2003.

[37] DataRay. DataRay Inc., Application Note, Beam Fit Algorithms, Rev. 0403A, 9 Mar. 2004. http://www.dataray.com, 2004.

[38] W. Demtröder. *Experimentalphysik 4, Kern-, Teilchen und Astrophysik (1. Auflage)*. Springer-Verlag, 1998.

[39] W. Demtröder. *Laser Spectroscopy, Basic Concepts and Instrumentation (3rd edition)*. Springer, 2003.

[40] W. Demtröder. *Experimentalphysik 2, Elektrizität und Optik (3. Auflage)*. Springer-Verlag, 2004.

[41] A. P. Van Devender and P. G. Kwiat. High Efficiency Single Photon Detection via Frequency Up-Conversion. *Journal of Modern Optics*, 51(9–10):1433–1445, 2004.

[42] M. Dine, W. Fischler, and M. Srednicki. A Simple Solution to the Strong CP Problem With A Harmless Axion. *Physics Letters B*, 104(3):199, 1981.

[43] C. Du, Z. Wang, J. Liu, X. Xu, B. Teng, and et al. Efficient intracavity second-harmonic generation at 1.06 μm in a BiB_3O_6 (BiBO) crystal. *Applied Physics B*, 73:215–217, 2001.

[44] K. Ehret, M. Frede, S. Ghazaryan, M. Hildebrandt, E.-A. Knabbe, D. Kracht, A. Lindner, J. List, T. Meier, N. Meyer, D. Notz, J. Redondo, A. Ringwald, G. Wiedemann, and B. Willke. Resonant laser power build-up in ALPS - A light shining through a wall experiment. *Nuclear Instruments and Methods in Physics Research A*, 612:83–96, 2009.

[45] K. Ehret, M. Frede, S. Ghazaryan, M. Hildebrandt, E.-A. Knabbe, D. Kracht, A. Lindner, J. List, T. Meier, N. Meyer, D. Notz, J. Redondo, A. Ringwald, G. Wiedemann, and B. Willke. New ALPS results on hidden sector lightweights. *Physics Letters B*, 689:149–155, 2010.

[46] M. Fouché, C. Robilliard, S. Faure, C. Rizzo, and et al. Search for photon oscillations into massive particles. *Physical Review D*, 78:032013, 2008.

[47] P. Franken, A. Hill, C. Peters, and G. Weinreich. Generation of Optical Harmonics. *Physical Review Letters*, 7:118, 1961.

Bibliography

[48] M. Frede, B. Schulz, R. Wilhelm, P. Kwee, F. Seifert, B. Willke, and D. Kracht. Fundamental mode, single-frequency laser amplifier for gravitational wave detectors. *Optics Express*, 15(2):459–465, 2007.

[49] J. A. Frieman, C. T. Hill, A. Stebbins, and I. Waga. Cosmology with Ultralight Pseudo Nambu-Goldstone Bosons. *Physical Review Letters*, 75(11):2077, 1995.

[50] P. Fritschel. Second generation instruments for the Laser Interferometer Gravitational Wave Observatory (LIGO). *Proc. SPIE*, 4856(22):282–291, 2003.

[51] Y. Furukawa, K. Kitamura, and et al. Green-induced infrared absorption in MgO doped $LiNbO_3$. *Applied Physics Letters*, 78(14):1970–1972, 2001.

[52] J. R. Gair, I. Mandel, A. Sesana, and A. Vecchio. Probing seed black holes using future gravitational-wave detectors. *Classical and Quantum Gravity*, 26(20):204009, 2009.

[53] L. Gallais, H. Krol, J. Y. Natoli, M. Commandré, M. Cathelinaud, L. Roussel, M. Lequime, and C. Amra. Laser damage resistance of silica thin films deposited by Electron Beam Deposition, Ion Assisted Deposition, Reactive Low Voltage Ion Plating and Dual Ion Beam Sputtering. *Thin Solid Films*, 515:3830–3836, 2006.

[54] H. Gies, J. Jaeckel, and A. Ringwald. Polarized Light Propagating in a Magnetic Field as a Probe for Millicharged Fermions. *Physical Review Letters*, 97:140402, 2006.

[55] M. Giovannini, E. Keihänen, and H. Kurki-Suonio. Big bang nucleosynthesis, matter-antimatter regions, extra relativistic species, and relic gravitational waves. *Physical Review D*, 66:043504, 2002.

[56] S. Greenstein and M. Rosenbluh. Dynamics of cw intra-cavity second harmonic generation by PPKTP. *Optics Communications*, 238:319–327, 2004.

[57] Nature Publishing Group. Not a WISP of evidence, 2010.

[58] H. E. Haber and G. L. Kane. The Search for Supersymmetry: Probing Physics Beyond the Standard Model. *Physics Reports*, 117:75–263, 1985.

[59] M. J. Hadley. Classical Dark Matter. *arXiv:gr-qc/0701100v1*, 2007.

[60] W. Heisenberg and H. Euler. Folgerungen aus der Diracschen Theorie des Positrons. *Zeitschrift für Physik*, 98:714, 1936. [arXiv:physics/0605038].

[61] J. S. Heyl and L. Hernquist. Birefringence and dichroism of the QED vacuum. *Journal of Physics A: Mathematical and General*, 30:6485–6492, 1997.

[62] S. Hild, H. Grote, J. Degallaix, and et al. DC-readout of a signal-recycled gravitational wave detector. *Classical and Quantum Gravity*, 26:055012, 2009.

[63] F. Hoogeveen and T. Ziegenhagen. Production and Detection of Light Bosons Using Optical Resonators. *Nuclear Physics B*, 358:3–26, 1991.

[64] D. S. Hum, R. K. Route, G. D. Miller, and et al. Optical properties and ferroelectric engineering of vapor-transport-equilibrated, near-stochiometric lithium tantalate for frequency conversion. *Journal of Applied Physics*, 101:093108, 2007.

[65] G. Hummelt. SOLID-STATE LASERS: TEM_{00} CW green laser source is a powerful tool. *Laser Focus World*, 42(8):77, 2006.

[66] J. Isern, S. Catalan, E. Garcia-Berro, and S. Torres. Axions and the white dwarf luminosity function. *Journal of Physics: Conference Series*, 172:012005, 2009.

[67] J. Jaeckel, J. Redondo, and A. Ringwald. Signatures of a Hidden Cosmic Microwave Background. *Physical Review Letters*, 101:131801, 2008.

[68] J. Jaeckel and A. Ringwald. The Low-Energy Frontier of Particle Physics. *Annual Review of Nuclear and Particle Science*, 60:405–437, 2010.

[69] J. H. Jang, I. H. Yoon, and C. S. Yoon. Cause and repair of optical damage in nonlinear optical crystals of BiB_3O_6. *Optical Materials*, 31:781–783, 2009.

Bibliography

[70] N. Jarosik, C. L. Bennett, J. Dunkley, B. Gold, M. R. Greason, and et al. Seven-Year Wilkinson Microwave Anisotropy Probe (WMAP) Observations: Sky Maps, Systematic Errors, and Basic Results. *The Astrophysical Journal Supplement Series*, 192:14, 2011.

[71] I. Juwiler and A. Arie. Efficient frequency doubling by a phase-compensated crystal in a semimonolithic cavity. *Applied Optics*, 42(36):7163, 2003.

[72] T. Kane and R. L. Byer. Monolithic, unidirectional single-mode Nd:YAG ring laser. *Optics Letters*, 10(2):65–67, 1985.

[73] M. Katz, R. K. Route, D. S. Hum, K. R. Parameswaran, G. D. Miller, and M. M. Fejer. Vapor-transport equilibrated near-stoichiometric lithium tantalatefor frequency-conversion applications. *Optics Letters*, 29(15):1775–1777, 2004.

[74] W. Koechner. *Solid-State Laser Engineering (5th edition)*. Springer-Verlag Berlin, 1999.

[75] H. Kogelnik and T. Li. Laser Beams and Resonators. *Applied Optics*, 5(10):1550, 1966.

[76] E. Komatsu, K. M. Smith, J. Dunkley, C. L. Bennett, B. Gold, G. Hinshaw, N. Jarosik, and et al. Seven-Year Wilkinson Microwave Anisotropy Probe (WMAP) Observations: Cosmological Interpretation. *The Astrophysical Journal Supplement Series*, 192:18, 2011.

[77] F. J. Kontur, I. Dajani, Y. Lu, and R. J. Knize. Frequency-doubling of a CW fiber laser using PPKTP, PPMgSLT, and PPMgLN. *Optics Express*, 15(20):12882–12889, 2007.

[78] S. C. Kumar, G. K. Samanta, and M. Ebrahim-Zadeh. High-power, single-frequency, continuous-wave second-harmonic-generation of ytterbium fiber lasers in PPKTP and MgO:sPPLT. *Optics Express*, 17(16):13771, 2009.

[79] P. Kwee. *Laser Characterization and Stabilization for Precision Interferometry*. Dissertation, Gottfried Wilhelm Leibniz Universität Hannover, 2010.

[80] P. Kwee, F. Seifert, B. Willke, and K. Danzmann. Laser beam quality and pointing measurement with an optical resonator. *Review of Scientific Instruments*, 78(7):073103, 2007.

[81] P. Kwee and B. Willke. Automatic laser beam characterization of monolithic Nd:YAG nonplanar ring lasers. *Applied Optics*, 47(32):6022, 2008.

[82] N. Lastzka and R. Schnabel. The Gouy phase shift in nonlinear interactions of waves. *Optics Express*, 15:7211–7217, 2007.

[83] Y.-S. Lee, T. Meade, V. Perlin, H. Winful, T. B. Norris, and A. Galvanauskas. Generation of narrow-band terahertz radiation via optical rectification of femtosecond pulses in periodically poled lithium niobate. *Applied Physics Letters*, 76(18):2505, 2000.

[84] Z. M. Liao, S. A. Payne, J. Dawson, A. Drobshoff, C. Ebbers, and D. Pennington. Thermally induced dephasing in periodically poled KTP frequency-doubling crystals. *Journal of the Optical Society of America B*, 21(12):2191, 2004.

[85] J. T. Lin, J. L. Montgomery, and K. Kato. Temperature-tuned noncritically phase-matched frequency-conversion in LiB3O5 crystal. *Optics Communications*, 80:159–165, 1990.

[86] Z. Lin, L. F. Xu, R. K. Li, Z. Wang, C. Chen, M. H. Lee, E. G. Wang, and D.-S. Wang. *Ab initio* study of the hygroscopic properties of borate crystals. *Physical Review B*, 70(23):233104, 2004.

[87] A. Lindner. The Future of Low Energy Photon Experiments. *arXiv:0910.1686*, 2009.

[88] A. Lindner. personal communication, 2011.

Bibliography

[89] V. Loriette and C. Boccara. Absorption of low-loss optical materials measured at 1064 nm by a position-modulated collinear photothermal detection technique. *Applied Optics*, 42:649–656, 2003.

[90] O. A. Louchev, N. E. Yu, S. Kurimura, and K. Kitamura. Thermal inhibition of high-power second-harmonic generation in periodically poled $LiNbO_3$ and $LiTaO_3$ crystals. *Applied Physics Letters*, 87:131101, 2005.

[91] T. H. Maiman. Stimulated Optical Radiation in Ruby. *Nature*, 187:493–494, 1960.

[92] G. Mann. *Experimentelle und theoretische Untersuchungen zur Frequenzkonversion von Nd:YAG-Laserstrahlung mit hoher Durchschnittsleistung*. Dissertation, Technische Universität Berlin, 2003.

[93] G Mann, S Seidel, and H Weber. Influence of mechanical stress on the conversion efficiency of KTP and LBO. In Tiziani, HJ and Rastogi, PK, editor, *Laser Metrology and Inspection*, volume 3823 of *Proceedings of the Society of Photo-Optical Instrumentation Engineers (SPIE)*, pages 289–297. SPIE - International Society of Photo-Optical Instrumentation Engineers, 1999. Conference on Laser Metrology and Inspection, MUNICH, GERMANY, JUN 14-15, 1999.

[94] J. D. Mansell, J. Hennawi, E. K. Gustafson, M. M. Fejer, R. L. Byer, and et al. Evaluating the effect of transmissive optic thermal lensing on laser beam quality with a Shack-Hartmann wave-front sensor. *Applied Optics*, 40(3):366–374, 2001.

[95] K. I. Martin, W. A. Clarkson, and D. C. Hanna. 3 W of single-frequency output at 532 nm by intracavity frequency doubling of a diode-bar-pumped Nd:YAG ring laser. *Optics Letters*, 21:875, 1996.

[96] L. McDonagh and R. Wallenstein. Low-noise 62 W CW intracavity-doubled TEM_{00} $Nd:YVO_4$ green laser pumped at 888 nm. *Optics Letters*, 32(7):802, 2007.

[97] T. Meier. Optical Resonators in Current and Future Experiments of the ALPS Collaboration. In David B. Tanner and Karl A. van Bibber, editors, *Axions 2010 - Proceedings of the International Conference*, volume 1274 of *AIP Conference Proceedings*, pages 156–162, 2010. http://link.aip.org/link/?APCPCS/1274/156/1 ; also published as *arXiv:1003.5867v1 [physics.optics]*.

[98] T. Meier, B. Willke, and K. Danzmann. Continuous-wave single-frequency 532 nm laser source emitting 130 W into the fundamental transversal mode. *Optics Letters*, 35(22):3742–3744, 2010.

[99] J. E. Midwinter and J. Warner. The effects of phase matching method and of uniaxial crystal symmetry on the polar distribution of second-order non-linear optical polarization. *British Journal of Applied Physics*, 16(8):1135, 1965.

[100] A. J. Miller, S. W. Nam, J. M. Martinis, and A. V. Sergienko. Demonstration of a low-noise near-infrared photon counter with multiphoton discrimination. *Applied Physics Letters*, 83:791, 2003.

[101] V. Mürk, V. Denks, A. Dudelzak, P. P. Proulx, and V. Vassiltsenko. Gray tracks in KTiOPO4: Mechanism of creation and bleaching. *Nuclear Instruments and Methods in Physics Research Section B: Beam Interactions with Materials and Atoms*, 141(1-4):472 – 476, 1998.

[102] G. Mueller, P. Sikivie, D. B. Tanner, and K. van Bibber. Detailed design of a resonantly enhanced axion-photon regeneration experiment. *Physical Review D*, 80:072004, 2009.

[103] D. N. Nikogosyan. Beta Barium Borate (BBO). *Applied Physics A*, 52:359–368, 1991.

[104] D. N. Nikogosyan. Lithium Triborate (LBO). *Applied Physics A*, 58:181–190, 1994.

[105] W. Nolting. *Grundkurs Theoretische Physik (6th edition)*. Springer-Verlag, 2002.

Bibliography

[106] R. Paschotta. *Encyclopedia of Laser Physics and Technology*. Wiley, 2008. also available at http://www.rp-photonics.com.

[107] R. D. Peccei. The Strong CP Problem and Axions. *Lecture Notes in Physics*, 741:3–17, 2008.

[108] R. D. Peccei. Why PQ? In David B. Tanner and Karl A. van Bibber, editors, *Axions 2010 - Proceedings of the International Conference*, volume 1274 of *AIP Conference Proceedings*, pages 7–13, 2010. http://link.aip.org/link/?APCPCS/1274/7/1.

[109] P. J. E. Peebles and B. Ratra. The cosmological constant and dark energy. *Reviews of Modern Physics*, 75:559–606, 2003.

[110] M. Peltz, J. Bartschke, A. Borsutzky, R. Wallenstein, S. Vernay, T. Salva, and D. Rytz. Harmonic generation in bismuth triborate (BiB_3O_6). *Applied Physics B*, 81:487–495, 2005.

[111] X. Peng, L. Xu, and A. Asundi. High-power efficient continuous-wave TEM_{00} intracavity frequency-doubled diode-pumped Nd:YLF laser. *Applied Optics*, 44(5):800, 2005.

[112] W. J. Percival, B. A. Reid, D. J. Eisenstein, N. A. Bahcall, and et al. Baryon acoustic oscillations in the Sloan Digital Sky Survey Data Release 7 galaxy sample. *Monthly Notices of the Royal Astronomical Society*, 401.2148–2168, 2010.

[113] M. E. Peskin and D. V. Schroeder. *An Introduction to Quantum Field Theory*. Westview Press, ABP, 1995.

[114] PI. *PrincetonInstruments*™ datasheet of CCD PIXIS 1024B (Rev. M.1). http://www.princetoninstruments.com/products/speccam/pixis/, 2010.

[115] M. Pierrou, F. Laurell, H. Karlsson, T. Kellner, C. Czeranowsky, and G. Huber. Generation of 740 mW of blue light by intracavity frequency doubling with a first-order quasi-phase-matched KTiOPO$_4$ crystal. *Optics Letters*, 24:205, 1999.

[116] M. Punturo, M. Abernathy, F. Acernese, B. Allen, N. Andersson, and et al. The Einstein Telescope: a third-generation gravitational wave observatory. *Classical and Quantum Gravity*, 27:194002, 2010.

[117] S. Rabien, R. I. Davies, T. Ott, S. Hippler, and U. Neumann. PARSEC, the Laser for the VLT. *Proceedings of SPIE*, 4494:325–335, 2002.

[118] J. Redondo. Bounds on very weakly interacting sub-ev particles (wisps) from cosmology and astrophysics. In A. Lindner, J. Redondo, and A. Ringwald, editors, *Proceedings of the 4th Patras Workshop on Axions, WIMPs and WISPs*, volume 1, pages 23–26. Verlag Deutsches Elektronen Synchrotron, 2008. 4th Patras Workshop on Axions, WIMPs and WISPs, June 18-21, 2008, Hamburg, Germany; available online at http://www-library.desy.de/confprocs.html.

[119] J. Redondo and A. Ringwald. Light shining through walls. *arXiv:1011.3741*, 2010.

[120] I. Ricciardi, M. De Rosa, A. Rocco, P. Ferraro, and P. De Natale. Cavity-enhanced generation of 6 W cw second-harmonic power at 532 nm in periodically-poled MgO:LiTaO$_3$. *Optics Express*, 18(11):10985, 2010.

[121] A. G. Riess, L. Macri, S. Casertano, M. Sosey, H. Lampeitl, and et al. A Redetermination of the Hubble Constant with the Hubble Space Telescope from a Differential Distance Ladder. *The Astrophysical Journal*, 699:539–563, 2009.

[122] M. Roncadelli, A. De Angelis, and O. Mansutti. A New Light Boson from Cherenkov Telescopes Observations? *Nuclear Physics B - Proceedings Supplements*, 188:49, 2009.

Bibliography

[123] J. Sakuma, Y. Asakawa, and M. Obara. Generation of 5-W deep-UV continuous-wave radiation at 266 nm by an external cavity with a $CsLiB_6O_{10}$ crystal. *Optics Letters*, 29(1):92, 2004.

[124] G. K. Samanta, S. C. Kumar, K. Devi, and M. Ebrahim-Zadeh. Multicrystal, continuous-wave, single-pass second-harmonic generation with 56% efficiency. *Optics Letters*, 35:3513–3515, 2010.

[125] G. K. Samanta, S. C. Kumar, M. Mathew, C. Canalias, V. Pasiskevicius, F. Laurell, and M. Ebrahim-Zadeh. High-power, continuous-wave, second harmonic generation at 532 nm in periodically poled $KTiOPO_4$. *Optics Letters*, 33(24):2955–2957, 2008.

[126] P. R. Saulson. *Fundamentals of interferometric gravitational wave detectors*. World Scientific, 1994.

[127] B. F. Schutz. Gravitational wave astronomy. *Classical and Quantum Gravity*, 16:A131–A156, 1999.

[128] S. J. Sheldon, L. V. Knight, and J. M. Thorne. Laser-induced thermal lens effect: a new theoretical model. *Applied Optics*, 21:1663–1669, 1982.

[129] A. Siegman. *Lasers (1st edition)*. University Science Books, 1986.

[130] P. Sikivie. Axion theory. In A. Lindner, J. Redondo, and A. Ringwald, editors, *Proceedings of the 4th Patras Workshop on Axions, WIMPs and WISPs*, volume 1, pages 13–16. Verlag Deutsches Elektronen Synchrotron, 2008. 4th Patras Workshop on Axions, WIMPs and WISPs, June 18-21, 2008, Hamburg, Germany; available online at http://www-library.desy.de/confprocs.html.

[131] P. Sikivie. Dark matter axions. *arXiv:0909.0949*, 2009.

[132] S. Sinha, D. S. Hum, K. E. Urbanek, Y.-W. Lee, M. J. F. Digonnet, M. M. Fejer, and R. L. Byer. Room-Temperature Stable Generation of 19 Watts of

Single-Frequency 532-nm Radiation in a Periodically Poled Lithium Tantalate Crystal. *Journal Of Lightwave Technology*, 26(24):3866, 2008.

[133] S. Somekh and A. Yariv. Phase Matching by Periodic Modulation of the Nonlinear Optical Properties. *Optics Communications*, 6(3):301–304, 1972.

[134] S. Spiekermann, F. Laurell, V. Pasiskevicius, H. Karlsson, and I. Freitag. Optimizing non-resonant frequency conversion in periodically poled media. *Applied Physics B*, 79:211, 2004.

[135] F. Steier, R. Fleddermann, J. Bogenstahl, and et. al. Construction of the LISA back-side fibre link interferometer prototype. *Classical and Quantum Gravity*, 26:175016, 2009.

[136] T. Südmeyer, Y. Imai, H. Masuda, N. Eguchi, M. Saito, and S. Kubota. Efficient 2nd and 4th harmonic generation of a single-frequency, continuous-wave fiber amplifier. *Optics Express*, 16(3):1546, 2008.

[137] L. R. Taylor, Y. Feng, and D B. Calia. 50 W CW visible laser source at 589 nm obtained via frequency doubling of three coherently combined narrow-band Raman fibre amplifiers. *Optics Express*, 18(8):8540, 2010.

[138] A. Thüring. *Investigations of coupled and Kerr non-linear optical resonators*. Dissertation, Gottfried Wilhelm Leibniz Universität Hannover, 2009.

[139] F. Torabi-Goudarzi and E. Riis et al. Efficient cw high-power frequency doubling in periodically poled KTP. *Optics Communications*, 227:389–403, 2003.

[140] S. V. Tovstonog, S. Kurimura, I. Suzuki, K. Takeno, S. Moriwaki, N. Ohmae, N. Mio, and T. Katagai. Thermal effects in high-power CW second harmonic generation in Mg-doped stochiometric lithium tantalate. *Optics Express*, 16(15):11294, 2008.

[141] T. Umeki, M. Asobe, Y. Nishida, O. Tadanaga, K. Magari, T. Yanagawa, and H. Suzuki. Highly Efficient +5-dB Parametric Gain Conversion Using

Bibliography

Direct-Bonded PPZnLN Ridge Waveguide. *IEEE Photonics Technology Letters*, 20(1):15, 2008.

[142] T. Umeki, O. Tadanaga, and M. Asobe. Highly Efficient Wavelength Converter Using Direct-Bonded PPZnLN Ridge Waveguide. *IEEE Journal of Quantum Electronics*, 46(8):1206, 2010.

[143] J. T. Verdeyen. *Laser electronics (3rd edition)*. Prentice Hall, 1995.

[144] J. Villarroel, M. Carrascosa, A. Garcia-Cabanes, and et al. Photorefractive response and optical damage of $LiNbO_3$ optical waveguides produced by swift heavy ion irradiation. *Applied Physics B*, 95:429–433, 2009.

[145] S. Wang, V. Pasiskevicius, and F. Laurell. Dynamics of green light-induced infrared absorption in $KTiOPO_4$ and periodically poled $KTiOPO_4$. *Journal of Applied Physics*, 96:2023, 2004.

[146] C. Wetterich. Cosmology and the fate of dilatation symmetry. *Nuclear Physics B*, 302:668–696, 1988.

[147] A. G. White, J. Mlynek, and S. Schiller. Cascaded second-order nonlinearity in an optical cavity. *Europhysics Letters*, 35:425–430, 1996.

[148] F. Wilczek. Axions and Family Symmetry Breaking. *Physical Review Letters*, 49(21):1549, 1982.

[149] B. Willke. Stabilized lasers for advanced gravitational wave detectors. *Laser And Photonics Reviews*, 4(0).780–794, 2010.

[150] B. Willke, P. Aufmuth, C. Aulbert, S. Babak, and et. al. The GEO 600 gravitational wave detector. *Classical and Quantum Gravitation*, 19:1377, 2002.

[151] B. Willke, K. Danzmann, M. Frede, P. King, D. Kracht, P. Kwee, O. Puncken, R. L. Savage, B. Schulz, F. Seifert, C. Veltkamp, S. Wagner, P. Wessels, and L. Winkelmann. Stabilized lasers for advanced gravitational wave detectors. *Classical and Quantum Gravity*, 25:114040, 2008.

[152] W. Winkler, K. Danzmann, A. Rüdiger, and R. Schilling. Heating by optical absorption and the performance of interferometric gravitational-wave detectors. *Physical Review A*, 44:7022–7036, 1991.

[153] S. T. Yang, C. C. Pohalski, E. K. Gustafson, R. L. Byer, R. S. Feigelson, and et al. 6.5-W, 532-nm radiation by cw resonant external-cavity second-harmonic generation of an 18-W Nd:YAG laser in LiB_3O_5. *Optics Letters*, 16(19):1493, 1991.

[154] J. Q. Yao, W. Q. Shi, J. E. Millerd, G. F. Xu, E. Garmire, and M. Birnbaum. Room-temperature 1.06-0.53-μm second-harmonic generation with $MgO:LiNbO_3$. *Optics Letters*, 15(23):1339, 1990.

[155] Y. K. Yap, T. Inoue, H. Sakai, Y. Kagebayashi, Y. Mori, and et al. Long-term operation of $CsLiB_6O_{10}$ at elevated crystal temperature. *Optics Letters*, 23:34, 1998.

[156] E. Zavattini, G. Zavattini, G. Ruoso, G. Raiteri, E. Polacco, E. Milotti, V. Lozza, M. Karuza, U. Gastaldi, G. Di Domenico, F. Della Valle, R. Cimino, S. Carusotto, G. Cantatore, and M. Bregant. New pvlas results and limits on magnetically induced optical rotation and ellipticity in vacuum. *Phys. Rev. D*, 77(3):032006, Feb 2008.

[157] I. Zawischa. personal communication, 2010.

[158] Y. Zheng, F. Li, Y. Wang, K. Zhang, and K. Peng. High-stability single-frequency green laser with a wedge $Nd:YVO_4$ as a polarizing beam splitter. *Optics Communications*, 283:309–312, 2010.

[159] K. Zioutas, M. Tsagri, Y. Semertzidis, T. Papaevangelou, T. Dafni, and V. Anastassopoulos. Axion Searches with Helioscopes and astrophysical signatures for axion(-like) particles. *New Journal of Physics*, 11:105020, 2009.

[160] F. Zwicky. Republication of: The redshift of extragalactic nebulae. *General Relativity and Gravitation*, 41(1):207–224, 2009.

Acknowledgements

My doctoral studies at the AEI Hannover were a great and interesting time, which I enjoyed a lot. I would like to thank Prof. Karsten Danzmann for giving me this opportunity to work in an institute with such a remarkable infrastructure. For the very personal mentoring, the relaxed atmosphere and for the large number of scientific discussions and helpful advices, I would like to thank Benno Willke.

I enjoyed the fruitful and cooperative daily work with the other (ex-)members of the laser group, namely Patrick Kwee, Jan Pöld, Robin Bähre, Christina Bogan, Henning Ryll, Marina Dehne, Michaela Pickenpack and Reza Hodajerdi. A special thanks goes to Frank Seifert, not only for teaching me electronics. I am grateful for all the small (non-)scientific talks with my colleagues at the AEI. These talks were often a source of valuable amusement and hence eased the daily work.

Furthermore, a good part of this thesis would not have been possible without the people of the ALPS collaboration. I would like to thank all of you for allowing me to become part of the group, and I would like to thank the DESY in general for providing the biggest part of the ALPS infrastructure. A special thanks goes to Axel Knabbe, with whom I enjoyed to work on the experiment.

Now, at the very end, I would like to switch back to my mother tongue.

An dieser Stelle möchte ich meinen Eltern und meinem Bruder herzlich für all die Unterstützung in den vergangenen Jahrzehnten danken. Schließlich richtet sich der letzte Satz dieses Werks an die Person, deren Unterstützung mir in den letzten Jahren die Wichtigste war – Tine, das hier richtet sich an Dich!

Die VDM Verlagsservicegesellschaft sucht für wissenschaftliche Verlage abgeschlossene und herausragende

Dissertationen, Habilitationen, Diplomarbeiten, Master Theses, Magisterarbeiten usw.

für die kostenlose Publikation als Fachbuch.

Sie verfügen über eine Arbeit, die hohen inhaltlichen und formalen Ansprüchen genügt, und haben Interesse an einer honorarvergüteten Publikation?

Dann senden Sie bitte erste Informationen über sich und Ihre Arbeit per Email an *info@vdm-vsg.de*.

Sie erhalten kurzfristig unser Feedback!

VDM Verlagsservicegesellschaft mbH
Dudweiler Landstr. 99
D - 66123 Saarbrücken
www.vdm-vsg.de

Telefon +49 681 3720 174
Fax +49 681 3720 1749

Die VDM Verlagsservicegesellschaft mbH vertritt

Printed by Books on Demand GmbH, Norderstedt / Germany